"This is a very beautiful, passionate collection of essays. If for Jacques Ellul *La Technique* was already 'the wager of the century' in 1954, when he wrote his famous book, in the twenty-first century it has become a nightmare! More than ever, the thought of Ellul helps us to analyze, understand, and act in the face of the propaganda for efficiency at any price, the political illusion, and ecological and climate change threats. By their contributions, the authors in this collection allow a new generation to discover what for me was an awakening of conscience and a guide for my activism."

—**José Bové**, former student of Jacques Ellul

"It's an educational pleasure to endorse this magnificent book. It benefits in coherence from the Berkeley conference, where its chapter authors were in conversation. Its competent scholarship makes *Political Illusion & Reality* one of the most valuable books I have ever read on the topic."

—**Cliff Christians**, emeritus professor of Media and Communications at the University of Illinois

"In my four years on the Davis City Council, Ellul accompanied me to the dais, and everywhere I went in my community. Ellul challenged me to confront the damaging spirit of 'technique' by creating local, face-to-face, and truly human responses to the challenges of my city. He cautioned me to consider the true ends of human flourishing and to not become enamored with efficient means without reference to those ends. Finally, he provided a language that helped me challenge the 'technicians' in places of power who 'do not accept [the] moral judgement' of their actions. This book is a reminder of the spiritual and moral grounding of this great man's work. Without him and his ideas, I could not have succeeded in helping guide my small city in these turbulent times."

—**Robb Davis**, former mayor and city council member in Davis, California

"Provocative. Informative. Enlightening. At a time when we place our trust in the political system and the state to meet all our needs and protect us from harm, Ellul challenges us to rethink the object of our trust and affections. Anyone seriously interested in understanding the place of government in our lives today will want to read this book."

—**Geraldine E. Forsberg**, instructor of media ecology at Western Washington University

"Jacques Ellul's insights on politics and faith are profound and, in our current crisis of political illusion, are needed more than ever. The essays in *Political Illusion & Reality* are a labor of love that will bear fruit as a helpful and inspiring guide through Ellul's thought."

—**Sharon Gallagher**, editor of *Radix Magazine*

"This stimulating and varied international collection of essays shows how the insights of Jacques Ellul on technique, politics, propaganda and so much else help us understand what is going on today beneath the headline political turmoil, and how in various contexts those inspired by Ellul are not just thinking in fresh ways, but acting to make a difference."

—**Andrew Goddard**, senior research fellow, Kirby Laing Institute for Christian Ethics

"*Political Illusion & Reality* provides hope for those discouraged by current cruel political realities. Drawing on Jacques Ellul's deep and illuminating insights, a new narrative is proposed here which allows an exit from traditional social contract clichés. The Jewish and Christian peaceable kingdom myth (Andy Alexis-Baker) enables 'ideological detoxification.' Practical essays—by activists like Sylvie Justome and mayors like Noel Mamere and Robb Davis—demonstrate the relevance of this perspective. This is a must-read for regaining spiritual and political courage!"

—**Carole J. Lambert**, professor of English at Azusa Pacific University

"The essays in this international collection can only help Jacques Ellul to be recognized, at last, as one of the great thinkers of our time. They do real justice to someone who anticipated the political and philosophical crises which stretch across the so-called developed world in the present. If we want to understand where we are today, we must read and reread Ellul. This book from an International Jacques Ellul Society colloquium at Berkeley, California, will help everyone to better grasp the universal relevance of Ellul's thought. Even in France, where we often quote the saying 'A prophet is not without honor save in his own country,' I hope this indispensable collection will also be translated and published in France."

—**Noël Mamère**, former mayor of Bègles, France, and former student of Jacques Ellul

"Once Ellul teaches you how to see the world, you will never view it the same way again. We are, therefore, fortunate to have *Political Illusion & Reality* to help us see Ellul for the extraordinary sociological and theological thinker he was. Be warned, however—seeing the way Ellul sees will not make your life easy."

—**Stanley Hauerwas**, emeritus professor of Ethics and Law at Duke University

"*Political Illusion & Reality* is a stunning collection of essays by a remarkable community of scholars influenced by Jacques Ellul. I was taken aback by its range and insightfulness, and its testimony to the continuing vitality of Ellul. I commend it to anyone who doubts the relevance of Ellul's sociological, philosophical, and theological analyses of the technological way of life within which we are all now increasingly enmeshed."

—**Carl Mitcham**, international professor of Philosophy of Technology at Renmin University of China in Beijing

"No two readers understand an author in exactly the same way. There are few authors like Jacques Ellul who write existentially, pushing us to confront our various beliefs, ideas, and lives. This stimulating collection of essays demonstrates this. No one will agree with each essay, but everyone will discover new insights into and new applications of Ellul's thought."

—**Richard Stivers**, distinguished professor of Sociology at Illinois State University

"The essays in *Political Illusion & Reality* show us why Jacques Ellul's prophetic analyses of politics, the state, and international relations still matter. Understanding the world can be a way of already beginning to change it. In these essays, Ellul's ideas 'out-narrate' the existing stories of what is going on and provide multiple concrete examples of how to appropriate and apply his insights."

—**Scott M. Thomas**, associate professor of International Relations at the University of Bath in the United Kingdom

"For years, Jacques Ellul was required reading in the Sojourners community. *Political Illusion & Reality* is a well-curated collection of essays and commentaries that reminds us of his continued timeliness, as well as the depth and complexity of thought that he has to offer us. Ellul's vision of political society is morally challenging, provocative, and constructive, and this edition has been edited in such a salient way for the times in which we now find ourselves."

—**Jim Wallis**, author of *America's Original Sin: Racism, White Privilege, and the Bridge to a New America*

Political Illusion and Reality

Political Illusion and Reality

Engaging the Prophetic Insights of Jacques Ellul

EDITED BY
David W. Gill
AND
David Lovekin

◂PICKWICK *Publications* · Eugene, Oregon

POLITICAL ILLUSION AND REALITY
Engaging the Prophetic Insights of Jacques Ellul

Copyright © 2018 Wipf and Stock Publishers. All rights reserved. Except for brief quotations in critical publications or reviews, no part of this book may be reproduced in any manner without prior written permission from the publisher. Write: Permissions, Wipf and Stock Publishers, 199 W. 8th Ave., Suite 3, Eugene, OR 97401.

Pickwick Publications
An Imprint of Wipf and Stock Publishers
199 W. 8th Ave., Suite 3
Eugene, OR 97401

www.wipfandstock.com

PAPERBACK ISBN: 978-1-5326-4906-6
HARDCOVER ISBN: 978-1-5326-4907-3
EBOOK ISBN: 978-1-5326-4908-0

Cataloguing-in-Publication data:

Names: Gill, David W., editor. | Lovekin, David, editor.

Title: Political illusion and reality : engaging the prophetic insights of Jacques Ellul / Edited by David W. Gill and David Lovekin.

Description: Eugene, OR: Pickwick Publications, 2018 | Includes bibliographical references.

Identifiers: ISBN 978-1-5326-4906-6 (paperback) | ISBN 978-1-5326-4907-3 (hardcover) | ISBN 978-1-5326-4908-0 (ebook)

Subjects: LCSH: Ellul, Jacques, 1912–1994. | Technology and civilization. | Civilization, Modern—1950–. | Political science.

Classification: LCC BX4827.E5 P7 2018 (print) | LCC BX4827.E5 (ebook)

Manufactured in the U.S.A.   10/04/18

For our colleagues, friends, and fellow-workers at

The Ellul Forum
for the Critique of Technological Civilization
(FOUNDED 1988)

and

The International Jacques Ellul Society
(FOUNDED 2000)

www.ellul.org

Contents

Preface by David Gill and David Lovekin | xi

Part One: **Foundations** | 1

1. Fascism, Son of Liberalism by Jacques Ellul | 3
   TRANSLATED AND INTRODUCED
   BY Jacob Rollison (ABERDEEN, SCOTLAND)

2. Bernard Charbonneau, the State, and Politics | 45
   BY Daniel Cérézuelle (BORDEAUX, FRANCE)

3. Oui et Non: Jacques Ellul's Dialectical Engagement with Karl Marx | 57
   BY Jacob Van Vleet (BERKELEY, CALIFORNIA)

4. The Political Theology of Jacques Ellul | 67
   BY David Gill (BERKELEY, CALIFORNIA)

5. Jacques Ellul and the Nonviolent Movements of His Time in France | 89
   BY Frédéric Rognon (STRASBOURG, FRANCE)

6. Liberalism and the State in French and Canadian Technocritical Discourses: Intersections and Contrasts between George Grant and the Bordeaux School | 104
   BY Christian Roy (MONTREAL, QUEBEC)

7. Jacques Ellul and Charles Taylor on the Sacrality of Secularism | 115
   BY Gregory Wagenfuhr (CANON CITY, COLORADO)

8. Illusion, Revolt, and Revolution | 126
   BY Samir Younés (NOTRE DAME, INDIANA)

9. Jacques Ellul, the Symbol, and the Political Illusion | 137
   BY David Lovekin (HASTINGS, NEBRASKA)

Part Two: **Applications** | 151

10. Resistance in Thought and Action: From the Global to the Local at Verdon | 153
    BY Sylvie Justome (BORDEAUX, FRANCE)

11. Political Upheavals Old and New: The Radical Vision of Jacques Ellul | 160
    BY Langdon Winner (TROY, NEW YORK)

12. Christian Political Engagement in a New Key? Reading Ellul in Ottawa | 169
    BY Paul C. Heidebrecht (OTTAWA, ONTARIO)

13. Bringing Ellul to the City Council: Mayor Robb Davis Reflects on Ellul's Influence | 181
    BY Mark D. Baker (FRESNO, CALIFORNIA)

14. Jacques Ellul & Thai Politics since 2000 | 192
    BY Virginia Landgraf (CHICAGO, ILLINOIS)

15. The Hashtag Comes First: Jacques Ellul and Occupy Wall Street | 207
    BY Zach Loeb (PHILADELPHIA, PENNSYLVANIA)

16. The Illusion of Change: From Greece to Italy and Spain | 219
    BY Carmen Madorrán (MADRID, SPAIN)

17. Ellul and Health Care Reform | 229
    BY Raymond Downing (ELDORET, KENYA)

18. The Environmental Movement in Ellulian Perspective | 239
    BY Jason Hudson (CALVER, DERBYSHIRE, ENGLAND)

19. Charlottesville vs. the Real Revolution | 251
    by Michael Morelli (aberdeen, scotland)

Part Three: **Appropriations** | 261

20. Democracy Confiscated? The Continued Relevance of Jacques Ellul | 263
    by Patrick Chastenet (bordeaux, france)

21. Sham Universe: Notes on the Disappearance of Reality in a World of Hallucinations | 270
    by Doug Hill (pasadena, california)

22. An Insight Into Ellul's Necessary Revolution | 279
    by Adrián Almazán Gómez (madrid, spain)

23. Telling a Better Story: Jacques Ellul Against Statist Mythology and For a New World | 285
    by Andy Alexis-Baker (chicago, illinois)

Preface

BY David Gill AND David Lovekin

How will our era be described by future generations? What will careful observers and researchers identify as the primary characteristics of our political and cultural life? For Americans, will the Barack Obama era be viewed as some kind of great leap forward (that just as quickly fell backward)? And how will future generations explain the presidency of Donald Trump? It hardly seems possible to find two greater opposites than Obama and Trump. But beneath their seemingly vast differences in governing philosophy, political practice, personal ethics, and character, were the commonalities more significant than the differences? Did the election results matter in any real way or was it all an illusion, full of sound and fury but signifying nothing—while the real action was the growth and impact of a technological state, fundamentally the same and in charge, not just under Obama and Trump, but under Putin, Merkel, and Xi Jinping? What is political illusion, and what is reality?

The author provoking our responses to such questions is Jacques Ellul (1912–1994). Ellul produced more than fifty books and a thousand articles in newspapers and assorted journals and yet regarded himself as an author of one long book with many chapters. Politics and government played large in this massive *oeuvre*. His studies in the history of law and social institutions, in theology and biblical criticism, and in social and ethical criticism clustered around the notion that technology, *la technique,* was not just machines and devices but a mentality, an ensemble of methods, a way of viewing the world, that had come to dominate political, social, cultural and intellectual life from the eighteenth century to the present.

Ellul's best known work, published in 1954 in France, *La Technique ou L'enjeu du siècle,* was translated as *The Technological Society* in 1964 with a strong recommendation by Aldous Huxley and with a glowing foreword by distinguished sociologist Robert K. Merton.[1] This book was read actively on many college campuses during the Sixties and Seventies when the "establishment" was called into question. Ellul demonstrated that the social bedrocks of economics and politics were marginalized by demands for efficiency and a rational methodology that used science and mathematics when expedient but rarely gave way to doubt and caution. Students often felt victimized by methodologies that reduced them to numbers, marks on lotteries for the Vietnam war, or to a one-size-fits-all grey flannel suit with a cubicle in a suburban office building.

For technique, more is always better than less and quantity is quality where "number" rules; choices are made automatically for the "efficient" as an absolute that is always on the horizon. The new is always the best until it is no longer new. This is not the physicist's efficiency but its opposite: more effort strives to achieve less taken as "more," which is the "efficient." "What can be done will be done," Ellul announced. Moral, aesthetic, or philosophical considerations were put aside.

Ellul claimed that politics became illusory when it was deemed the most important aspect of human life at a time when real politics no longer existed. Politics as the consideration of values that benefit the common good has given over to the means to provide technical services and support; technique itself becomes the common good. Political debate is silenced in a discourse seeking only unification that hovers around technology that has become the final arbiter. "Is it efficient?" is the question that silences dispute. So what we think is meaningful politics is really mostly illusion: it is an illusion to think we can control the state and direct it, an illusion to think there are political solutions to most of our human challenges and needs. Politics generates great passions and emotions but it is in the realm of the "ephemeral," the transitory, the surface, the fleeting, the insubstantial. Meanwhile the bureaucratic technological state grinds on and grows out, assisted by technicized propaganda and public relations in a world of images.

* * * *

In July of 2016, sixty-five scholars, pastors, activists, writers, artists, and other Ellul-readers from North America and Europe gathered for three days on the campus of the University of California, Berkeley, to consider "Politics without illusion, revolution without violence: revisiting the thought of

---

1. Jacques Ellul, *The Technological Society,* trans. John Wilkinson (New York: Alfred A. Knopf, 1964).

Jacques Ellul." Fifty years earlier Berkeley was in the vanguard of a radical questioning and protest against what had then become ordinary politics. No reason to stop now, especially with the shocking successes of the disruptive revolt of Trump and his movement in 2016!

We (Lovekin and Gill) were part of the conference organizing team and yielded to many requests to follow up with a book collecting the conference papers. Most of the twenty-three chapters in this book are revised and expanded versions of what were conference presentations in July 2016. However we broadened our call for papers and that resulted in a few additions. The most important of those additions, and the longest chapter in this book, was Jacob Rollison's translation (for the first time into English) of Ellul's own 1936 article "Fascism, son of Liberalism."

It is frankly amazing to realize that Ellul was only twenty-four years old when he wrote this. Remember also that by 1936 the specter of Nazism and fascism was on the rise and threatening civilization. By 1940 France was occupied by the Germans, Ellul had been fired from his university post, his father imprisoned, and he had joined the Resistance. To reflect on what in Liberalism (in the classic political sense, not in the Fox News epithet sense) could (or did) give birth to Fascism is an essential if painful task. With the rise of quasi-fascist right wing movements in the USA, France, Italy, Germany, the Philippines, and many other places, it is urgent that we return to this analysis.

Our book leads off with Ellul's essay and translator Jacob Rollison's helpful introduction. That essay and the following eight chapters comprise Part One, "Foundations." We want to understand Ellul's political thought more fully and clearly and view it in relation to his influences such as Bernard Charbonneau, Marx, and the Bible, to some of the political movements of his time, and to contemporary thinkers such as George Grant and Charles Taylor.

In Part Two, we present reflections on ten different "Applications" of Ellul's thought ranging from (1) specific *topics* such as health care and ecology to (2) particular *events* or *movements* (Occupy Wall Street, the recent Charlottesville conflict, Sixties radicalism, Syriza, Podemos, Five Star, the Verdon (France) protest, etc.), to (3) particular political *contexts* (Thailand, Ottawa, and the Davis (California) mayor's office). Some of these reflections are by scholars but other observations are by engaged participants and activists in the thick of things. The writing styles vary accordingly as we want them to speak in their own voices.

Finally in Part Three, "Appropriations," we close with four essays drawing on Ellul to speak to our current situation. From a young graduate student in Madrid (Almazan), to a veteran New York and Los

Angeles-based journalist (Hill), to the University of Bordeaux's Professor of Political Science and the leading Ellul scholar in the world (Patrick Chastenet), to an Ellul-style anarchist and Loyola-Chicago university professor (Alexis-Baker)—we are all challenged to a serious and deep awareness of our own time and place—to be followed by a responsible resistance to evil and an incarnation of the good.

* * * *

We (Gill and Lovekin) want first to express our gratitude to the authors and translators represented in this volume. They (and we) do not do this for the money or fame! Second, we thank those whose behind-the-scenes generosity in time, effort, and finance made the conference as well as this book possible (including Lucia Gill, Terry Lovekin, Mark and Megan Kvamme, Randy Ataide, Mark Mayhle, Clay Radke, and our indefatigable IJES Executive Director, Ted Lewis). We are also especially grateful to Jerome Ellul (grandson of Jacques) for his enthusiastic partnership in our conferences and publication projects.

We dedicate this book to two related organizations: first, the *Ellul Forum for the Critique of Technological Civilization*. Founded in 1988, this semi-annual journal has for thirty years been the central communication channel for Ellul scholars, students, and readers. Second, we dedicate the book to the International Jacques Ellul Society, founded in 2000, alongside its francophone sister-society the Association Internationale Jacques Ellul. Both the journal and the society are labors-of-love kept operational by small donations and membership dues and countless hours of volunteer labor by a modest-size community of Ellulians around the world. We salute the *Ellul Forum* and the IJES and wish them many more decades of service and success.

David W. Gill, Berkeley, California
David Lovekin, Hastings, Nebraska
January 2018

Part One: **Foundations**

THE ESSAYS IN PART One are foundational to understanding Jacques Ellul's political thought, its sources and major themes. There is no substitute for a direct reading of Ellul's *The Technological Society*, especially its ninety-page fourth chapter on "Technique and the State," and *The Political Illusion*. Nevertheless, published here in English translation for the first time, Ellul's long and challenging 1936 article "Fascism, Son of Liberalism," written when he was just twenty-four years of age, is a powerful statement of some core themes in Ellul's political thought. Jacob Rollison provides the translation and a helpful introduction in chapter one.

In the four chapters that follow Ellul's, our authors reflect on important direct influences on Jacques Ellul's political thought. Daniel Cérézuelle describes Ellul's life-long friendship and dialogue with Bernard Charbonneau. Jacob Van Vleet writes about the influence of Karl Marx on Ellul. David Gill explores the biblically-grounded political theology he sees in Ellul. And Frederic Rognon describes Ellul's thought and career as it interacted with three significant non-violent movements of his time.

Two chapters then help us understand Ellul by way of comparative study, first with George Grant as analyzed by Christian Roy, then with Charles Taylor as analyzed by Greg Wagenfuhr. Finally, Samir Younes explores Ellul's perspectives on revolution and revolt in the face of the political illusion—and David Lovekin offers a reflection on symbol and boundary as a way to grasp the heart of Ellul's political message.

1

Fascism, Son of Liberalism BY Jacques Ellul

TRANSLATED & INTRODUCED BY Jacob Rollison

Jacques Ellul (1912–1994) lived almost his entire life in Bordeaux in southwestern France. From 1944 until his retirement in 1980 he was Professor of the History & Sociology of Institutions in the Faculty of Law & Economic Sciences at the University of Bordeaux, where he also held a Chair in the Institute of Political Studies. He was author of nearly sixty published books translated into many languages. He was one of the first and most powerful analysts and critics of the growth and impact of "technique" (not just the obvious technological machines and devices but the processes, methods, and values at the core of technology) on every aspect of human life and in every corner of the world. Ellul not only wrote many books but over one thousand articles, essays, and reviews that continue to influence, instruct, and inspire today.

Jacob Rollison is a doctoral student in theological ethics at the University of Aberdeen, Scotland. He holds a BA in Economics from Wheaton College (Illinois) and an MA in Media & Communication from the European Graduate School in Saas-Fee, Switzerland. He is the author of Revolution of Necessity: Language, Technique, and Freedom in Jacques Ellul and Slavoj Žižek *(Atropos Press, 2016).*

* * * *

Introduction by Jacob Rollison

JACQUES ELLUL'S 1936 ARTICLE "Fascism, Son of Liberalism," one of the first of over one thousand articles he would publish in his lifetime, provides a succinct and comprehensive introduction to Ellul's early political and sociological thought.[1] It showcases his impressive command of the historical evolution of western institutions, his capacity for profound insight and prophetic commentary, his inheritance and modifications of the thought of Karl Marx, the brutal realism of his approach to current events characteristic of his "Christian realism," his will for a revolutionary solution, and the sociological roots of his concern with propaganda. At the same time, the dense prose also indicates its status as an early piece, as does the directness of its concluding call for a response from the reader. I will discuss these two characteristics of the article below after giving some historical context which can help us understand the article's true significance.

Historical Context

It is astounding to consider that Ellul wrote this article at only twenty-four years of age. By this time, Ellul was already an intellectual of considerable power. With his father unemployed and the family impoverished on the eve of the 1929 economic crisis, a teenage Ellul was giving lessons in Latin, Greek, German, and French to ten-year olds.[2] After becoming a *bachelier* at age sixteen (roughly the equivalent of a U.S. high school diploma), by eighteen Ellul discovered the thought of Karl Marx and Søren Kierkegaard and would go on to read everything written by both.[3] During this period, he also experienced what he described as a "brutal" conversion to Christianity.[4] By age twenty-four, he finished his doctorate in law. His studies were in the history and evolution of law as practiced in western societies (writing his dissertation on an ancient Roman legal institution), rather than studying to practice as a courtroom lawyer. In 1937, Ellul received his first teaching post at the University of Montpellier.

Ellul was beginning his teaching and writing career on the eve of World War II. The following year, he was given a post at the University of Strasbourg in Alsace (a region historically contested between France and Germany). As Germany's expansionist aspirations grew, the Strasbourg faculty was evacuated to Clermont-Ferrand in central France in 1939, and

---

1. Rognon, *Pensée*, 25.

2. Ellul, *In Season*, 9.

3. Rognon, *Pensée*, 24–25, 171, and 211. Rognon suggests that Ellul discovered Marx in late 1929 or early 1930, that his conversion to Christianity took place in August 1930, and that he discovered Kierkegaard within the following months.

4. Ellul briefly describes his conversion, about which he rarely spoke, in Ellul and Troude-Chastenet, *Conversations*, 52.

Ellul was soon fired from his post for comments against the new authoritarian government based in nearby Vichy (which collaborated with the Nazi regime after the 1940 armistice).

Earlier, in 1935, out of curiosity, Ellul had attended a Nazi meeting, which directly inspired his concerns and later work on propaganda. In an interview, he described how this event revealed to him the fearful power of a crowd: "Ellul: 'It was impressive to see how a crowd could be so easily galvanized, unified . . . nobody had any personal reactions left.' *Chastenet: 'And you, did you get caught up in the crowd reaction at that instant?'* Ellul: 'No, but the difficulty was to not raise my hand at the same time as the others.'"[5]

Ellul saw himself as an average person, whose reactions were common and ordinary, and he often worked by analyzing his own experiences and reactions.[6] Thus in writing this article, Ellul, with an uncommonly broad and deep knowledge of the historical evolution of western institutions, is reacting to what he sees as a significant shift, perhaps without precedent in western history, in the scale and scope of the power which politics can possess over a crowd of average people—likely, because he was deeply frightened by his own reaction as an outsider in a Nazi rally crowd.[7] For Ellul, as we will see, fascism can be seen as just an attempt to sustain liberal democracy under new conditions of propaganda.

Ellul's recognition of this problem and his search for a human response did not take place in a vacuum, however. Ellul forged his view of society in constant dialogue with his close friend Bernard Charbonneau, and both were involved in the French Personalist movements of the 1930s. Responding to a society in crisis, these movements advocated restructuring and decentralizing government decision-making. In their view, large-scale, centralized government ruled by abstract processes which followed the criteria of mathematical efficiency, treating people as manageable by a series of calculations. By contrast, they wanted to re-subordinate these processes to local, human decisions made for personal criteria, to restore to politics human concerns and a human scale.

---

5. Ellul and Troude-Chastenet, *Conversations*, 63. I have modified the English translation with reference to the original, in Ellul and Chastenet, *À contre-courant: Entretiens*, 137.

6. Ellul, *In Season*, 189–90.

7. Another Barthian Protestant thinker who was prominent in the personalist movement, Denis de Rougemont, famously reported an almost identical experience, dated March 11, 1936, in his notes from a stay at the University of Frankfurt as French reader over the winter of 1935–36, *Journal d'Allemagne* (reissued by the same publisher in the collected diaries of his *Journal d'une époque 1926–1946* in 1968).

In "Fascism, Son of Liberalism," we will see Ellul engaging with historical and contemporary issues and movements specific to his region. Prominent among these was the review *Esprit*, led by Catholic thinker Emmanuel Mounier.[8] Ellul and Charbonneau initially worked within the *Esprit* movement, hoping to give it a distinctive revolutionary character. They later parted ways, chiefly over *Esprit*'s liberal mentality and organization, and a concomitant willingness to view politics as one technique among others. These ways of thinking allowed them to entertain notions of collective personhood which Ellul and Charbonneau rejected (in Ellul's case, for theological as well as sociopolitical reasons). These features of Mounier's increasingly mainstream "communitarian personalism" would be confirmed by *Esprit*'s ambiguous evolution, turning the Christian personalism in which Ellul and Charbonneau participated into a fellow-travelling adjunct to a number of successive trendy ideologies, from the Popular Front through the Vichy regime's National Revolution to communism. These movements shared technocratic modernization as their common denominator, which was virtually unchallenged in post-war France when Ellul published his first books.

"Fascism, Son of Liberalism" was initially a lecture given to the Bordeaux personalist group, as part of a series of talks by Ellul and Charbonneau around *Esprit*'s personalist manifesto. It was printed over the course of 1936 in their local personalist newsletter, where the same theme of the pre-fascist sentiments inherent to liberal society had been discussed since 1935.[9] It was then taken up as an article by *Esprit* itself in its February 1937 issue. Ironically, this text reached the whole movement toward the end of Ellul's engagement with it, as a prophetic indictment of broad social trends from which it proved hardly immune.[10]

In this context, we can appreciate the essay's providential timing. It appeared before the full atrocities of World War II, during a time of political polarization and economic crisis, amid the rise of the fascist dictatorships

---

8. del Bayle, "Aux origines de la pensée de Jacques Ellul?," 34.

9. Roy, "Aux sources," 67–100, especially 77, 82–83, on the genesis and context of Ellul's text. For an overview in English of the distinctive place of Charbonneau and Ellul among the various strands of the French personalist movement, see Roy, "Ecological Personalism," 33–44.

10. In his article "Le Personnalisme et Mounier," 6–7, Ellul notes that the *Esprit* movement degenerated between 1937 and 1939. Rognon notes that Ellul split with *Esprit* later in 1937 (Rognon, *Pensée*, 25); Roy points to evidence of late March 1938 as the date of Charbonneau's final break with Mounier in the name of the personalist groups he effectively led with Ellul in Southwestern France, irrespective of their formal allegiances to the Paris-based reviews *Esprit* or *L'Ordre Nouveau* (Roy, "Aux sources," 89–90).

of Benito Mussolini in Italy and Adolf Hitler in Germany and against the backdrop of Stalin's Marxist despotism in Russia. And within this context, it provides a uniquely perceptive critique not only of these particular authoritarian regimes, but also of the liberal or progressive ones claiming to oppose them, even as they blithely stand on the same perilous and compromised sociological ground.

Ellul's Sociopolitical Thought

But to stop at lauding Ellul's resistance would be to miss his point. What he is addressing is not limited to Hitler's Germany or Mussolini's Italy. Rather, he is describing is a fundamental shift in the way western political institutions function due to the power and application of new techniques, a shift which is the logical consequence of an evolution, not a random fluke. In other words, these dictatorial regimes are not mere accidents of twentieth century history. In fact, Ellul will argue that after World War Two, western nations unconsciously imitate these regimes in fundamental respects. To better grasp this institutional evolution, we need to review Ellul's use of Marx's thought and the perspective given by his historical study of western institutions.

"Fascism, Son of Liberalism" encapsulates Ellul's Marxist analysis of the twentieth century west, as well as his modifications to Marx's thought. Ellul admired Marx's global view of society and its evolution, which helped him make sense of his father's unemployment and the economic crises they were living through. He emulated this global view in his own thinking. Adapting Hegel's dialectics of history, Marx thought that history evolved via a play of tensions, and that careful sociological examination could find the most powerful of these tensions at a given moment, the "determining factor." In his own age, Marx identified this as economics, and he discussed it under the term *capital*. Ellul thought that Marx was correct for the nineteenth century, but that things had changed: in the mid-twentieth century, the new determining factor was *technique*. In Marx's era, economic production was largely a function of labor: more production required more labor. Marx's theory of the class struggle and the proletarian working class as revolutionary relied heavily on the fact that labor was so central to production, and that the workers were alienated since they did not own the product of their labour.

But by the mid-twentieth century, technological change meant that production could increase exponentially while labor could remain the same or even decrease. Ellul argues that by this point workers are now even happy to accept their alienation, which sterilizes their revolutionary potential. Thus, if the economy was the determining factor in Marx's time, for Ellul, technique now determines the economy (and thus socio-historical evolution) and the tensions between the classes conform to the new situation. Marx's hope for a revolutionary working class has been thwarted (at least temporarily). We can thus see that when Ellul decries the overemphasis on the economy as a closed system and the ignorance of the seriousness of technique as a factor of institutional evolution, he is both drawing on and modifying Marx's thought.[11] Marx had hoped that class tensions would surmount capitalism and result in a more equitable society; Ellul argues that technique has transformed liberal politics with a capitalist economy into a fascism with a command economy.

We can better understand Ellul's concern with technique if we remember that Ellul's primary academic training was in the history of institutions, focusing on the social evolution of the West from antiquity to the nineteenth century. His largest work is an untranslated five-volume history of institutions through this time period. In "Fascism, Son of Liberalism" Ellul is especially concerned with the ways fascism will transform *law*. This matters because for Ellul, law is at the heart of institutions, and institutions connect the individual *person* with the group.[12] Ellul saw a common pattern of juridical evolution across many societies, in which natural law was first perceived as divine and eternal. Then law becomes secularized, arising out of the will of the people. At a third stage, it becomes *technical* law, less spontaneous, more rigid, and more abstract. It is "technical" because the relation between all the laws as a system of law becomes more important than the relation between law and human life. Ellul often uses a phrase from Roman law, *summum jus, summum injuria*, to summarize this: the more perfect law becomes as a system, the more it actually excludes justice.[13]

---

11. This analysis is brief and oversimplified. For fuller treatments of Ellul on Marx, see Jacob Van Vleet's discussions in this volume; Menninger, "Marx in the Social Thought," chapter 1; or my own "Hope, Prophecy and Prediction." In French, see part 2.A.II of Rognon, *Pensée*. Most of Ellul's works containing his explicitly Marxist discussions are untranslated, but among translated works, see Ellul, *Autopsy*, reprint ed., for his critiques of Marx on revolution.

12. See Ellul, "Le Personnalisme, Révolution immédiate," a lecture given in October 1935 whose text appeared the next month in the first issue of the local internal newsletter of Friends of *Esprit*, initially called *Journal intérieur du groupe de Bordeaux*, and was reissued in *Cahiers*, 81–94.

13. This evolution of law is summarized in Ellul, *Theological Foundation*, chapter 1. With the tendency of this fascist (and more broadly, technical) law to repressive control, we could certainly see Ellul as a precursor to the works of Michel Foucault, a comparison that has been made on several occasions.

In this light, we can see why Ellul saw such a critical mutation in the move from liberalism to fascism. In fascism, law is no longer at the heart of society, but is "added onto it" as a façade. But if law had served as the distinctively *personal* link between the human and the group, what will now replace it? Essentially, technique, and *propaganda* as a technique of communications, becomes "the very means of government." The group is no longer linked in a process of discussion mediated by their own thought, but rather by common sentiments externally produced in various mass settings.[14] We are no longer a society of conscious human *people*, but an unconscious and manipulable *mass*. Thus, Ellul's personal fear at his reaction to the powerful propaganda of a Nazi rally, when he analyzes it, becomes a symptom of the transition from a society based on law to a society based on the use of powerful tools—that is, a *technological society*.[15]

Rhetorical Communication

I have also said that "Fascism, Son of Liberalism" is a piece of rhetorical communication. By this, I mean that its genre is ultimately *communication*, or rhetoric. Ellul is not writing a textbook, but addressing a public. While his essay contains economic analyses, Marxist debates, and philosophical references, all of this is meant as material to cause his readers to *become aware* of their situation in the world and respond to it. "Becoming conscious" was a major theme for both Kierkegaard and Marx (though differently for each), Ellul's intellectual ancestors. Only this "consciousness" keeps us from being a "concrete mass," from acting as a direct result of mass phenomena, and understanding our situation at the personal level is precisely what personalism was after. Ellul's essay was an attempt to become personally conscious in his own time; part of its value today is as critical inspiration, a model pushing us to understand our own times.

Ellul saw himself as following Kierkegaard in aiming to enact the *opposite* of this propagandizing mode of communication. He is addressing those in the mass and calling them to become individuals, inspiring them to free, personal action and resistance to social currents. Those of us who gather to discuss Ellul's life and works can attest to this experience. No

14. For a more recent, Ellul-influenced treatment of this "democracy of emotion," see Virilio, *Administration of Fear*.

15. This is the less-than-fully accurate English translation of the title of Ellul's first major book analyzing our society as an expression of technique—see Ellul, *Technological Society*. A more literal translation of the French would be something like "Technique, or the Stake of the Century." It must be noted that the English "technology" is not quite the same as what Ellul intends by *technique*, though this latter partially includes it.

two "Ellulians" are completely alike.[16] We are a community of singular, distinctive individuals, with commonalities to be sure, but sometimes very radical differences.

Lastly, "Fascism, Son of Liberalism" is a call to *contemplative work*. Ellul's task is a communicative one with a lot to communicate. He is trying to render a vast body of knowledge accessible and usable by the average intellectual. But for this to work, he calls the intellectual to a lot of work. His nearly sixty books are meant to be read alongside each other. Practically, this means that it is hard to have the proper context to understand what Ellul means until you have read, say, fifteen or so of his books.[17] The world Ellul describes can seem so hopeless that a reader lacking the balanced tension of his broader corpus might wrongly decide that violent revolt is the only way out (as Ellul reader and "Unabomber" Ted Kaczynski infamously demonstrated). Without the hope most directly described in his theological studies, Ellul once worried, his sociological critique could produce despair or even suicide.[18] Positively, Ellul calls for a true revolution that will change *us* before it will change anything else. As he wrote elsewhere, "the essence of revolution is an ethical relationship, consideration for others, and the acknowledgment of another man's (anyone's, and not just a select person's) rights and dignity."[19] In our era of blatant propaganda from political (and corporate and religious) leaders of all stripes, to say nothing of masses of social media manipulators, the mirror which Ellul's "Fascism, Son of Liberalism" gives the reader is a humbling call to self-examination, for fascism only appears when it is asked for.

Bibliography

de Rougemont, Denis. *Journal d'Allemagne*. Paris: Gallimard, 1938; reissued in *Journal d'une époque 1926–1946*. Paris: Gallimard, 1968.
Ellul, Jacques. *Autopsy of Revolution*. 1st ed. New York: Alfred A. Knopf, 1971.
———. *Autopsy of Revolution*. Reprint, Eugene, OR: Wipf & Stock, 2012.
———. *In Season, Out of Season: An Introduction to the Thought of Jacques Ellul*. Translated by Lani K. Niles. Based on interviews by Madeleine Garrigou-Lagrange. San Francisco: Harper & Row, 1982.
———. "Le Personnalisme et Mounier: Pourquoi je me suis séparé de Mounier." *Réforme* 265 (April 1950) 6–7.

16. For an extensive catalogue of such individuals, see Rognon, *Générations*.

17. See Greenman, Schuchardt, and Toly, *Understanding Jacques Ellul*, 146, for a similar suggestion.

18. Ellul and Chastenet, *Conversations*, 22.

19. Ellul, *Autopsy*, 292, 1st ed.

———. "Le Personnalisme, Révolution immediate." In *Cahiers Jacques Ellul: Les annees personnalistes* 1 (2003) 81–94.
———. *The Technological Society*. New York: Vintage, 1964.
———. *The Theological Foundation of Law*. London: SCM, 1961.
Ellul, Jacques, and Patrick Chastenet. *À contre-courant: Entretiens avec Jacques Ellul*. Paris: La Table Ronde, 2014 [1994].
———. *Jacques Ellul on Politics, Technology, and Christianity: Conversations*. Translated by Joan Mendes France. Eugene, OR: Wipf & Stock, 2005.
Greenman, Jeffrey P., Read Mercer Schuchardt, and Noah J. Toly. *Understanding Jacques Ellul*. Eugene, OR: Cascade, 2012.
Loubet del Bayle, Jean-Louis. "Aux origines de la pensée de Jacques Ellul? Technique et Société dans la réflexion des mouvements personnalistes des années 30." In *Cahiers Jacques Ellul: Pour une critique de la société technicienne, no. 1, Les années personnalistes*. Bordeaux: Pixagram, 2004.
Menninger, David C. "Marx in the Social Thought of Jacques Ellul." In *Jacques Ellul: Interpretive Essays*, edited by Clifford G. Christians and Jay M. Van Hook, 17–32. Chicago: University of Illinois Press, 1981.
Rognon, Frédéric. *Générations Ellul: soixante héritiers de la pensée de Jacques Ellul*. Geneva: Labor et Fides, 2012.
———. *Jacques Ellul: Une pensée en dialogue*. 2nd ed. Geneva: Labor et Fides, 2013.
Rollison, Jacob. "Hope, Prophecy and Prediction in Jacques Ellul's Christian Realism." In *Driven By Hope: Economics and Theology in Dialogue*, edited by Patrick Nullens et al., 95–106. Leuven: Peeters, 2017.
Roy, Christian. "Aux sources de l'écologie politique: Le personnalisme gascon de Bernard Charbonneau et Jacques Ellul." *Canadian Journal of History* 27 (April 1992) 67–100. https://roychristian.academia.edu/.
———. "Ecological Personalism: The Bordeaux School of Bernard Charbonneau and Jacques Ellul." *Ethical Perspectives* 6 (April 1999) 33–44. https://roychristian.academia.edu/.
Virilio, Paul. *The Administration of Fear*. Los Angeles: Semiotext(e), 2012.

Fascism, Son of Liberalism[1]

BY Jacques Ellul

Fascism: A Feigned Reaction

FASCISM APPEARS, WHEN VIEWED from the outside, as a reaction. This doctrine is so commonly accepted and so evident that nobody questions it: fascism is not a necessity, it is not an inevitable product of the modern world, but it is a reaction against this world. We invoke so-called historical laws defining history as a succession of actions and reactions, where each new institution would be explained by its opposite and explain the birth of its own negation. A kind of oversimplified Hegelian dialectic, if you will—this is a common idea.

But if we leave these generalities, once we try to specify this "opposite," we begin to disagree, depending on whether we are considering the state of affairs which preceded fascism, or the tendencies of its doctrine. Fascism becomes a reaction against liberalism in the first case, a reaction against communism in the second.

The first of these oppositions is especially drawn by the fascists themselves. They freely admit that their doctrine is not essentially positive. They proclaim that liberalism has given all that it could give—that it is no more

---

1. This article is a translation of Ellul, "Le fascisme, fils du libéralisme," 113–37. This text was first circulated in typed form in the mimeographed internal newsletter of the local group "Friends of *Esprit*," the *Bulletin du groupe de Bordeaux des Amis d'Esprit*, no. 4, then as an article in *Esprit* 53 (Feb. 1 1937) 761–97. Special thanks are due to Patrick Chastenet for publishing permissions, to Frédéric Rognon for help with several difficult phrases, and to Christian Roy for his careful and thorough review and editorial work on the entire text. All footnotes are by the translator and editors. Ellul's rhetorical style in this text tends to have less clarity than much of his later work, and so this translation has tried to stay strictly faithful to his words, punctuation, and tempo where possible. Where I found the original ambiguous, I have tried to communicate this ambiguity in the text, leaving any clarification in the footnotes.

than a dead doctrine in a world close to death and that we should take from this cadaver the elements which can serve us and order them according to a new spirit. It is a crude opposition to liberalism. It is enough for liberalism to have stated something for fascism to immediately proclaim the opposite—and these contrary statements are then piled up and presented as a body of doctrine. Fascism, having proclaimed the bankruptcy of decadent byzantinisms, and classified under this name[2] everything to do with real intelligence, is only one contradiction away from them. And nobody is shocked to see insurance companies developing in these countries where the values of risk are extolled, to see a gigantic bureaucratic apparatus where various Führers[3] proclaim their hatred of everything to do with administration, and to hear talk of the primitive and wild man, of the pure blond barbarian by means of the radio. All this to demonstrate beyond any doubt that if one talks of the barbarian, of risk, and of pen-pushers, it is only to counter a liberalism in which one spoke of Edmond Rostand,[4] sorbets, and comfort—a merely formal opposition between two sets of value which are equally sentimental and reductive.

What we should see as specific to fascism then, if we insist on seeing in it a reaction, is the formal *will* to reaction that it asserts against liberalism, and not reaction in a true sense. It wants to react, not only because it is carried by a current of public sentimentality, but also because it is imbued with the idea pointed out above that everything happens by action and reaction. What I propose to do here is precisely to investigate whether, through its forms, words, and expressions, it genuinely stands in opposition to liberalism or if there is a continuous current, an effective fusion of liberalism into fascism, and what has provoked this fusion.

To this superficial view of fascism by itself, as a reaction against liberalism, the communists oppose another no less superficial view, that of fascism as a reaction against communism. Fascism would thus appear as a weapon of decadent capitalism which, by this means, sought first to conquer communism, then to artificially regain strength. A kind of backwards mafia. We find here all the outdated notions of a world poorly known and

2. "Byzantinism" in this polemical context, but also proverbially, refers to irrelevantly complex discussions, which the sophisticated theological debates of ancient Eastern Christianity were often taken to be by modern Western opinion, hence the dismissive by-word "Byzantine."

3. *Führer* is the German word for "leader," made famous outside that language as the title chosen by Adolf Hitler as NSDAP leader and eventually dictator, on the model of *Duce* for Benito Mussolini, a fancier Italian word for the same concept, which other authoritarian leaders would emulate through the twentieth century.

4. Edmond Rostand (1868–1918) was a French writer famous for his play *Cyrano de Bergerac*.

poorly understood, these fictions that the parties of the left constantly stir up—the capitalist crouched in the wings, who makes the puppets move on the stage, while he, knowing all the weaknesses, all the ins and outs, all the means, seeks to make money—a primitive conception which supposes all too wily capitalists, capable of maneuvering with precision forces which can hardly be controlled.

To see in this enormous movement, which fascism represents, the product of a few backroom deals is really to oversimplify the issue. The world is perhaps a bit more complex than that. It may well be that capitalists' interests are served by it, though that is not absolutely certain. That they would finance fascist movements because they are afraid of communists, this is quite probable. But to believe that between them they have thought up a vast plan to renovate capitalism, and to believe that they have generated this movement from scratch is to disregard a lot of data, leaving aside the whole human aspect of the question, which is by far the most important. It is above all to forget that if money can finance such a movement, a necessity, a sort of fate was calling for it, an economic, political, ideological fate which made possible at a given moment this influence of money, wherever it came from, in whatever direction it would spread. Of course, if we insist on deciding between systems solely according to economic criteria, fascism will be classified among the capitalist systems; but we must not neglect the fact that it is established according to methods, on bases, with means and an aim which it holds in common with communism. Communism, too, is a formal negation of liberalism—and perhaps it, too, is its son. (To discuss this would lead us too far afield.)

Fascist Doctrine Comes After the Fact of Fascism

"Everything anti-liberal is ours!" fascism proclaims. Fascism is an act of defense, says communism. Mussolini wrote to Bianchi on August 27, 1921: "Right now, under pain of death, or worse, suicide, Italian fascism needs to provide itself with a body of doctrine. This expression is a little strong, but I would like it if the philosophy of fascism were created before the two months which separate us from the National Congress."[5] Now fascism had already been in existence for four years. Drawing these three statements together,

---

5. Benito Mussolini (1883–1945) was the leader of the Italian National Fascist Party, which rose to power in the aftermath of World War I; Mussolini ruled Italy as prime minister from 1922 to 1943, mostly as a dictator. Michele Bianchi (1883–1930) was one of the founders of the Fascist Party and influential in its regime, widely seen as a leader of its left wing.

we see—and this is a considerable observation on which we will dwell—we see that *fascist doctrine is only an outer element of fascism.* It comes to be added on to it, gives it a facade, glosses over it, but is not the very necessity of fascism. Fascism is born, it is a movement—better yet, a tendency, an exaltation which leads to the movement—only when it is launched. As it needs, on the one hand, to build bridges towards intellectuality, which is the foundation of the regime which precedes it, and on the other hand, to harmonize the various aspirations which appear, a decision is taken to create, within two months, a body of doctrine. Without this, suicide.

Fascism, then, would never appear, as brutal force sometimes does, to be conditioned by thought. It does not push brusquely into reality after having been long matured and prepared. It calls on feeling and not on intelligence; it is not an effort towards a real order but towards a fictional order of reality. It is preceded by a whole current of tendencies towards fascism. In all these countries we find these measures of policing and violence, this desire to curb the laws of parliament in the government's favor, statutory law and full powers, a systematic panic obtained by a slow pressure of newspapers on the common mentality, attacks against all dissident thought and expression, the limitation on freedom of speech and the right of assembly, the restriction of the right to strike and protest, etc. All these *de facto* measures already constitute fascism. They are the expression in reality of a state that fascism will do nothing but stabilize and legalize.

But this state is not admissible unless some prior preparation has come into play to form minds. This is the formation of a pre-fascist mentality. In short, we can consider that the *establishment of fascism happens thus: creation of a pre-fascist mentality . . . taking of fascist measures . . . Fascism . . . creation of a doctrine.* Of course, I cannot emphasize strongly enough that the first two phases are unaware of their fascist character. The pre-fascist mentality is made by itself, under the influences of the times. It is not a deliberate and subtle preparation to which Machiavellian schemers would subject these minds. It is made slowly because everyone listens to the same discourse, because everybody thinks of some impossible escape from the world where he lives, because everyone is fed on myths and the ideal, because people are in search of a better balance by the sacrifice of all which impedes it, because people want to renounce their real responsibility, their real risk, their real thought in favour of a proclamation of responsibility, of a will to risk, of a simulacrum of common thought—all destined to hide lacks and gaps. People are then ready to accept the leader.

What may help one grasp the reversal that I am proposing here (*namely, that the state of mind calls for fascism, and not a doctrine prior to a state of mind*) is the following fact: *the leader is born when fascism has become necessary.* Mussolini appears when the time is ripe, and if it weren't Mussolini, any general or industrialist would have carried the affair. The leader only comes into the world because the general mentality of the public demands this leader, calls for this hero in whom it wants to incarnate itself. Fascism is not a creation of the leader; the leader is a creation of the pre-fascist mentality.

The leader is there as it were to concretize the sometimes still unknown aspirations of the crowd—and this is what must be understood when I will speak of the demagoguery of fascism. It is not a question of a man who wants a world of such a fashion or of such a measure—*but of a man who strives to gather in himself all the commonplaces that the crowd accepts*, who catalogues all the virtues that the public demands and who thereby acquires a power, an influence over it. A common state of mind prior to fascism is a *sine qua non* condition of fascism. It is born of a certain complexity of the world. Before a situation which is more and more difficult, the crowd first follows those who were considered leaders until that point: the intellectuals. Now, the intellectuals betray us, and the best among them can say, at most, that the forces unleashed are so unforeseen, so unlimited, so unprecedented, that they do not understand much of them, that everything must be considered anew from the bottom up and that for the moment the path is dark.

The crowd does not like these admissions of powerlessness and does not like darkness. It prefers magicians who give perhaps the same admission, but wrapped in silver paper. And fascism has played on this. Not being able to explain, it has presented itself as a doctrine of hopelessness. There again, incidentally, it perfectly meets the state of mind of the average bourgeois, for whom it is a very remarkable attitude to be hopeless. Except that, while the intellectual of good quality offers him a genuine reason to despair, offers him good quality hopelessness, on the other side he is offered romantic hopelessness. All that is precise inspires fear because it demands an equally precise investigation and solution; what is precise is binding on the individual to the degree of its precision.

Fascism, being destined to express exactly the desire of a crowd, could not offer it an optimist doctrine since this crowd was drawn to pessimism, not only by a taste for thrills, but still more by the sense of latent crisis. Neither could it explain to the crowd the reasons to despair. This would have assumed that the crowd could understand, and for that matter, it would have had to be unpleasantly precise. And so, it portrayed itself as a pessimist doctrine: "all is lost, except through fascism; we have no more faith in saints nor in the apostles, we have no more faith in happiness nor in salvation; everything is going badly—and everything should go badly; we should leave material happiness to vile materialists, man should live from the ideal and not from bread; everything is in decline, culture and civilization, we must nevertheless fight to establish an order where these decadent cultures and civilisations would be banished." And it is always pleasant to reconstruct an order on new bases, even if we do not really know what they are. But we should be aware, given the importance of this common mentality which fascism secretes, *that this is possible in all countries: we cannot say that we will never allow this oppression in France*, or that in England fascism is foreign to tradition. These elements which form the pre-fascist mentality, like the style of Le Corbusier, are found to be identical in all countries.[6]

I will not insist anymore on this phenomenon of the creation of the pre-fascist mentality. This mentality, as I have said, tends to induce the acceptance of a number of authoritarian measures, for it is an abdication, and when these authoritarian measures are coordinated and complete, fascism is created. Nowhere have we seen the prior or decisive intervention of a doctrine. And indeed, there is no fascist doctrine. This explains very well the simultaneously primitive and terribly intellectual character of fascism's assertions. Completely separating fact and idea, it severs them in an even sharper demarcation than liberalism. Every idea is added on to the fact—we will return to this point in detail—all the rationalizations of fascist intellectuals to justify and explain fascism are never more than speculations on commonplaces—the very commonplaces that the crowd demands—to which it totally and willingly submits. Either old notions like the common good are taken up again in an essentially liberal formulation, or extravagant doctrines like the glorification of primitive man are added on.

It is thus quite evident that if we want to grasp fascism in its reality, we need not look for it in the constructs of intellectuals; it might be possible to proceed thus with communism, but fascism resists this by its very nature. To discuss the value of work or of the totalitarian state on the bases which Rocco or Villari offer us is to waste our breath, to work uselessly.[7] Fascism *is not to be studied in its doctrine because it is not a doctrine; it is*

---

6. Le Corbusier was the Swiss-born French architect Charles-Édouard Jeanneret (1887–1965), who took this pseudonym. An iconic originator (along with the Bauhaus) of the International Style of modern architecture, Le Corbusier designed low-cost, cubist, urban modular housing for the community of Pessac near Bordeaux (where Ellul would live for many years after the war) at the end of the 1920s, the Quartiers Modernes Frugès. In the context of this article's description of the creation of a new working class with a manufactured and standardized way of life and thought, Ellul is likely suggesting that a culture of such standardization is an unwitting liberal preparation for a transition to fascist society.

7. Alfredo Rocco (1875–1935), Italian politician and jurist whose theory of corporatism was later adapted to the ideology of the Fascist Party, and Luigi Villari, author of *The Fascist Experiment*.

a fact, produced by concrete historical situations. It is devoid of interest to discuss the various social forms of fascism, or, in a pure thesis, to oppose fascism to liberalism or to communism, because there are forces which go beyond these words, leading from one situation to the next. To study it, do not take free doctrinaires who attach it to Sorel or to Spengler, but statistics, and the cold description of a technical organization.[8] We must separate fascism from all ideas because in reality it is thus separated. We will see that it has perfected this final scission of thought and act, that it has utilized it. If, therefore, I am studying the passage from liberalism to fascism, I will do so only at the level of facts, from the angle of the economy, of political organization, of the community, etc..

From the Primacy of the Ideal to the Primacy of Method

Nevertheless, it is undeniable that, up to a point, fascism should be envisaged *from the perspective of its ideology*. A grand gesture is made and a magic word uttered to replace the absent doctrine: Enthusiasm, says the Colonel; *Fede*, says the Duce, *Wirkung*, says the Führer![9] And yet, people demand a faith in something, in *postulates*.[10] Fascism sets forth postulates that must be realized, and it is the study of these postulates that can have some interest. This is, first of all, because they are directly inspired by the average mentality and, secondly, because they express in a clear fashion the goal proposed by fascism. There is no contradiction between these two functions: the proposed goal is merely a more complete and more precise expression of what the crowd demands. A fascism draining the Pontine Marshes is greater than a Republic which has not cleared the Camargue.[11]

---

8. Georges Sorel (1847–1922), a French philosopher whose work inspired anarchists, Marxists, and fascists, among others, and who notably viewed violence as necessary for revolution, and Oswald Spengler (1880–1936), a German philosopher of history, who wrote *The Decline of the West* (originally published in 1918; most recently published by Forgotten Books in 2017). For Ellul's treatment of Sorel, see Ellul, *Les successeurs de Marx*, 54–66.

9. *Fede*: Italian for "faith" or "belief," a slogan used by Mussolini; *Wirkung*, a German word with many translations, including "action" and "effect," used as a slogan by Hitler; "the Colonel" refers to Colonel François de la Rocque (1885–1946) who, over the course of the 1930s, turned the Croix-de-Feu ("Fire Crosses") nationalist veterans' league into a right-wing mass movement that eventually became the Parti Social Français; for *Führer* and *Duce*, see note 3.

10. Postulates or axioms are statements that are simply asserted and assumed to be true.

11. This cryptic historical remark compares a swampy region south of Rome, dried and cultivated under Mussolini between 1928 and 1932, and the Camargue, a region

We will thus conduct a short exegesis of fascism's commonplaces, without setting out on a doctrinal study *per se*.

But first, I want to bring out a feature that I consider essential. *Fascism's lack of a theory is a liberal characteristic.* It is a consequence of liberalism. Throughout the period of liberalism, doctrines sprang up in large numbers. Never before had there been so many useless theories, so many competing and mutually contradictory systems. There were several reasons for this. First of all, freedom of thought—this is obvious. From the moment that there is a separation between thought and its consequences, the normal brake which used to rank the value of different thoughts disappears. There is no more direct repercussion for any thought expressed. *There is no longer any limit to the expression of thought.* Any thought that is hatched will just as quickly be expressed.

An obvious symptom of this problem is when a survey is made to find out if there is a crisis related to the book or a crisis in French thought. The endpoint of this crazy evolution is that what is in print is identified with thought. Morand is put on the same level as Bergson.[12] Discussion of the abstract, in the abstract, a confusion of thought and imagination. Someone who thought, knowing that for this act he would be brought to justice and perhaps be condemned to death, would still make a distinction in his thought between what was necessary and what was contingent; one does not risk one's neck for something contingent. The real and precise coming to consciousness of the power of thought by the one who thinks it is made incalculably more difficult by the fact that this thought no longer has any repercussion on his person, first of all, and then because it is lost in floods of books. *No discrimination is made anymore between the urgent and the unreal* because the urgent has itself become unreal. One no longer has any more consequences than the other, and the proclamation of a truth has no

---

in southeastern France known for its wetlands and western Europe's largest river delta (of the Rhone river).

12. Paul Morand (1888–1976), cosmopolitan French writer of witty short stories and travel novellas; Henri Bergson (1859–1941), French philosopher. Ellul is here taking up the gist of his friend Bernard Charbonneau's ironic critique of the quantification of French influence in terms of the book trade (putting in the same basket pulp entertainment and serious thought) in "Du prestige moral," the second and last installment of their short-lived "Exégèse des lieux communs" column for the *Bulletin de liaison des groupes "Ordre Nouveau"* 13 (July 1936) 2; that personalist movement's internal newsletter for its local cells. It was Ellul who had written the first one, making fun of the appeal to "personality everywhere" in advertising ("Personnalité partout" 12 (June 1, 1936) 4). Charbonneau had taken the same example of French thought being "hurt" by Italy's import ban on printed matter to illustrate the exegesis of commonplaces as the research method specific to personalism, by early 1936 in no. 2 of the *Bulletin du groupe de Bordeaux & des Amis d'Esprit*.

more importance than whatever is hatched by imagination. *By proclaiming freedom of thought, liberal society has freed itself from thought.* A constricted thought is always a dangerous power—abandoned to the four winds, it consumes itself in vain. This is why theories have multiplied without, for all that, society deviating one whit from its course.

The second reason for this multiplication is *our era's economic development. The material world tends to be organized on bases that are absolutely independent of any effort of thought.* The modern world tends to find in itself not only its own end, but also the reason for its development. It is ordained to a new principle, industrial technique, which makes its way into all human areas and tends to exclude everything that could trouble the strict play of its rules, its laws; in this case, it is thought which is excluded. *It thus appears necessary that thought remain separate from material development, that it be confined to the realm of abstraction* (of the crudest kind, as it happens). For it remains alien, in any form other than mathematical thought, to the rigorous and universal mastery of things that economic development implies.

The most striking example is that of political economy. As soon as it ceases to observe facts, it becomes a terrifying reality, all the more terrifying as it is applied to the very development of the things of which I have been speaking. A generality which stems from an abuse of logic, completely separated from facts, of countless abstractions (above all, the abstraction of the *homo economicus*), a refusal of contact with the concrete other than through statistics and regulations, the creation of airtight intellectual classifications, etc. This mental predisposition entailed by the proliferation of the modern economy was made worse by a morbid tendency to intellectual games, due to the fact that *intelligence, detached from the economic, moreover expatriated from existence, no longer had any necessity exterior to itself.* It could assuage all its desires, all its wild ideas. Machines would still continue to produce and the organization of a certain abundance would still arise. There was thus a *monopolization of intellectuality by the people who were assured of sufficient income, whatever their intellectual position might be.* Thus, in addition to the social, even legal risk, which was suppressed, economic risk was also suppressed for a class which was becoming at once the cultivated class and the owning class.

Amidst the abundance of theories which proliferated in the nineteenth century, we thus see three features of liberal thought emerge. First, *any thought is equivalent to any other thought*, no thought has dominant value, since none is constrained by action. None is urgent and necessary—all are contingent with respect to the order which is being established. Second, *any thought is admissible* since it is enough that it be justified intellectually by its coherence or its elegance alone. Third, *no theory has any chance of being realized*, and if it is necessary to move towards such a realization, nevertheless only reformism is admissible (as a consequence of the monopoly indicated above).

But there was a danger in this scission. Thought was glorified as never before. It was like heaven itself, a triumph of understanding as universal as brotherhood. It was tender and calm liberalism, full to the brim. But this thought was becoming incapable of readjusting to action. As long as action proved unneeded, as long as the world could keep turning all by itself, nobody noticed anything. But *this economic order which was thus made, ineluctable, inevitable, outside of human will and thought, ended up stumbling upon itself and no longer functioned very well.*

Later on, it was noticed that it no longer worked at all. *It was becoming necessary to act. But no doctrine* was made, no thought was ready, and distraught young intellectuals either refused to dirty their hands outside surrealism, or they denied purely and simply the influence of disorder on their thought, of which it was still a product, to be sure. All the old doctrines appeared identically abstract, equally valid and useless. The world could be reconstructed from a postulate, but this was useless for living. What was lost was the discrimination between thoughts, between those that are alive and those that are dead. Still, it was necessary to act, and yet, under pain of acting like fools, it was necessary to act with a semblance of reason, of coordination. What was needed was something immediately applicable to action and yet of higher origin than this action. In the face of thought disembodied from its role, there was now only one cry: "death to irrelevantly complex discussions—we must act."

To act, methods were found: it was no longer a reason to act that was sought, but only a justification for action. Doctrine was replaced by method—the electoral program. One could create a method for taking power just as much as a method for the resorption of surplus wheat, but no general thought would dominate or center the act. And thus, we see appear in the *realm of intelligence, the primacy of technique*, for method is nothing other than a technique of the intelligence. There again, technique triumphs over the human. Now this *passage from system to method exactly characterizes, from an intellectual point of view, the passage from liberalism to fascism*. There is a very direct link of parentage from one filiation to the other.

The liberal intellectual perversion, its intellectual treason, necessarily entails the turn towards a strict rule which will be codified, certified by fascism. It thus completes the radical scission between thought and life. This latter is enslaved to certain methods and certain techniques which must rigorously direct it. Incidentally, and *as long as life is in no way disturbed by it*, intelligence keeps all its value and the goddess. Thought is maintained in a high position, on a throne of clouds. Thus Goering, in line with pure liberal tradition, will say: "Achieve your salvation as you see fit," and Mussolini will write, "In the fascist State religion is considered one of the deepest manifestations of the human spirit: that is why it must not only be respected, but defended and protected."[13] The liberal State has slowly killed, by uselessness, by equality, by the all-too-tempting play which intellectuals are ever expected to indulge in, all power of thought. The fascist state has built the Pantheon where it has gathered these various cadavers, to which we still burn our incense, knowing they are no longer to be feared.

Liberal-Fascist Commonplaces

We now need only do a brief exegesis of the commonplaces of fascism to show that *fascism and liberalism are really using the same dead gods*. The same formulas are common for both.

We begin with *spirituality*. Our two supposedly opposed doctrines have exactly the same conception of it, and if they do not invoke exactly the same values, they both invoke them and do so with the same goal. We find here, on the same bases, the *contradiction between practical materialism and a spirituality of justification or of attitude*—one might say "of necessity" if this was not liable to cause a confusion between formal and real necessity.[14] Just as liberal spirituality demanded *a faith in reason*, and from there moved to call for only an abstract faith, so *fascism proclaims a revolt against science, a revolt against matter*, a quest for happiness in sacrifice, etc.. But in both cases, it is really what is material that is the foundation of life. Guizot proclaims: "Become richer!"[15] But it is not through allegories that we will idealize commerce and industry. The engravings of Luc-Olivier Merson and the paintings of Puvis de Chavannes express very well this hiding of a reality

---

13. Hermann Goering (1893–1946), a powerful NSDAP official who, among other things, oversaw Germany's air force from 1935 onward.

14. Ellul is alluding to different concepts of necessity developed by Hegel, and avoiding such language to prevent a complex Hegelian rabbit trail—and I will follow suit. The interested reader can find a discussion of necessity in Hegel's *Science of Logic*, vol. 1, sec. 3.

15. François Guizot (1787–1874), a doctrinaire classical liberal thinker, historian, and politician, whose heavy-handed resistance to the extension of the limited property-based franchise as Prime Minister of France led to the downfall of "citizen-king" Louis-Philippe's constitutional monarchy in the 1848 revolution that ushered in the short-lived Second Republic. "*Enrichissez-vous!*" was Guizot's infamous counsel to those who craved a broader electoral franchise, and has remained a byword for liberalism's shallowness and callousness in France, accounting in part for this ideology's relative discredit in that country's political culture.

that we basically admire, but which is not noble, behind an ideal image in cut wood, meant to ennoble it in an abstract fashion.[16] And opposite this, speeches about faith delivered standing on a tank, and Mussolini taking part in harvest festivals. There is no difference at all.

The cult of the primitive is itself but the normal and logical consequence of liberalism. Liberalism leads to an ever more frantic quest for whatever is novel. In the flood of accepted ideas and things, ever more prized and ever more abundant at the heart of a society where the intellectual is now only seen as an elegant and perfumed pariah, the intellectuals, who sense their uselessness, who feel they have become ancillary phenomena among human phenomena, can only acquire prestige by becoming spiteful critics of this society. If they push further than these useless invectives, they end up as cursed poets. The others are but university professors who preciously conserve this culture in their card indexes. *As a self-involved new caste, the intellectual feels tempted to seek the rare and the difficult*, whatever can be known only by the initiated. Henceforth, the artist will feel incapable of creating in this mediocre framework where he feels ill at ease because he feels useless. He will spend periods of far-off introspection in a darkened room, or he will leave for the Sunda Islands to bring back canvases and books that were unknown before him.[17]

Exoticism is born of this inability to really live in a world where everything repels you, which is no longer on your scale and which you no longer dominate. Consequently, all refinements are permitted and even recommended. One-upmanship in refinement flourished around 1900, but it resulted (since refinement, in the sense of thinning out, cannot be eternal) in a new focus on primitive arts, customs and cults. Just as a skilled poet pauses to make a cadence more evident, just like dissonance in harmony, so these *refinements extolled the cult of strength and the cult of spontaneity*. People went into ecstatic raptures about the moral value of Negro brass sections and the spirituality of hot jazz. Those who were incapable of spontaneity and strength were thrilled by spontaneity and strength as a foil to their refinement, as definitive proof of their understanding and

---

16. Merson (1846–1920) was a French academic painter and illustrator who did major decorative commissions for such institutions as the Palais de Justice. Pierre Puvis de Chavannes (1824–1898) was best known as "the painter for France" for his Symbolist murals in institutions such as the Sorbonne and Paris City Hall.

17. Possibly a reference to artists such as Paul Gauguin (1848–1903), who was celebrated for art resulting from visits not to the Sunda Islands, but to Tahiti and other islands in French Polynesia; and writer Robert Louis Stevenson (1850–1894), who travelled extensively through the islands of the Pacific.

perhaps, for that matter, since not all of them were radically perverted, as regret for a paradise lost.

Only something else was needed other than this desolation. Real action, which the world made impossible, was needed. This spontaneity needed to be lived, not described in scholarly tomes. Now there were philosophers who elevated this cult into a canon, giving it theoretical foundations. Was this a philosophy? It matters not. What I know is that this was to strength and the primitive roughly what Hugo's *The Hunchback of Notre Dame* was to the Middle Ages.[18] But this had an eminent quality. It represented a fictional thought of the era, a desire, a useless but definite tendency, and fascism seized this to concretize this thought in a sense of its own, to give to this useless tendency an all-too evident efficacy. *The desire for adventure was hijacked.* It was put into boots, made to march in step, made to witness beheadings with an axe and sworn to that it was thereby fulfilled. *The taste for the primitive was captured.* It was given garden parties, work camps were organized, there were choruses of spontaneous songs, violent speeches were made: this is what is called getting in touch with the concrete in our era.[19]

Finally, within the ideology of fascism, I will also single out *the defense of morality.* This is yet another specifically liberal fact. I am not saying, of course, morality in itself, the illustration of morality. I am referring to its verbal defense and justification. It is a well-known fact that the more a spiritual value is in decay, the more the language which expresses it becomes rigorous. To use Marxist terminology, it is as though the more the antagonisms between productive forces develop, the more the ideology of the privileged class becomes hypocritical. Thus, the more everyday life betrays the lie of words and common language, the more language will become sublime and virtuous. It is precisely a phenomenon of this kind that we are witnessing.

For liberalism, the moral act is essentially indifferent. As long as it is "understandable," the act does not call for judgement. And we have seen what an abstract machine this "understanding" has become. The act, which is not good or bad in itself, exists, and hence can be justified. *From the moral point of view, all acts have become abstract* in the liberal perspective, just as from a real point of view, all thought had become abstract. *But by this very fact, the moral law has been glorified even more, and it appears in the guise of a certificate of good conduct and character and of a duty to conform.*

Liberalism left things in this state, but fascism intervened, always in the same direction, with the essential role of crystallizing precisely this

---

18. A famous novel by French author Victor Hugo (1802–1885).

19. In French, "garden parties" is literally *fêtes champêtres*, a formal garden party popular in eighteenth-century France.

glorification in detached thought and encouraging morality and the sense of decency for the German race, as Killinger says.[20] And yet, the use of narcotics is common among fascist leaders, this being but the result of that. In this whole ideology, I will highlight but one more contradiction, because it is common to both ideologies. What is the point of changing ideologies if it fails, at least, to eliminate the contradictions?! It has to do with the general conception of life. It is the same liberals who praised the duty of collaboration and the struggle for life. It is the same fascists who speak of duties toward our fellows and of life as struggle. Formulas, yes, but what else is there beside formulas in all these ideologies? This contradiction of formulas is perfectly explained by the calls to heroism and to freedom on the one hand, by the recognition of a common interest and the superiority of the State on the other.

There is nothing original in fascist proclamations. We will see further the importance that they grant to the notion of the common good. But it is curious to find this notion covered in parade clothes. On the one hand, black clothes and top hats: freedom that we demand for individuals, provided that this freedom does not harm the common good, provided that it goes in the direction of the community, and provided that it observes the rules. On the other hand, rapiers and helmet feathers: the heroism that is expressed in shouts and outstretched arms, provided that it doesn't disturb order, that it is not the heroism of a single person but the heroism wanted by the State, provided that it observes the code of honour. In both cases, people proclaim that life is a fight but everyone knows that, in both cases, the swords are made of cardboard, the outcome of the fight is as well arranged, once and for all, as a theatrical play, and woe to whoever would break from this social determinism!

I will not insist any more on this ideological descent of fascism from liberalism. I have chosen very varied phenomena which are applicable to common facts of life. Let us move on to more material questions. Since fascism is more of a fact than a doctrine, I am actually going to study it first *in the economy*. I see two elements of kinship between the liberal economy and the fascist economy: *the technique of production, and trade unions.*

---

20. Manfred Freiherr von Killinger (1886–1944), a Nazi politician. Ellul's insight finds ultimate confirmation in Vanessa Lapa's *The Decent One*, a 2014 documentary that uses a cache of letters, diaries, and documents to follow the life of SS leader Heinrich Himmler in his own words, as a banal product of early twentieth century national-liberal common decency.

The Fascist Economy as Crystallization of the Restrictive Liberal Economy

The liberal economy was obsessed with the question of production. It had to produce as much as possible, and in doing so, it had to develop what was called the general economy. Liberalism insisted on the fact that the best method of production was, without question, the method of free competition and of free trade. But speculation was made on precise reasoning. The ever-growing production capacities were taken into account from the technical point of view, but only in the past, that is, the current state of production was taken to be definitive. It was thus a matter of *finding the system that would have made higher production economically possible*, or, if not higher, at least cheaper economically, and only economically. It was the play of economic forces that was calculated and not that of technical forces. From time to time, statistics could deceive, but not for long. At most, they served to bewilder the pessimist liberalism of those who promised starvation in the short term. The failure was due first of all to the fact that, in its calculations, the economy was based on an abstract man whose needs and reactions it was looking for in the absolute. It thought it could quantify this "nature," and it drew up charts of figures for human needs and utilities, enacting in a decisive fashion the transmutation of the qualitative into the quantitative.

Therein lies the second error of the liberal economy. It wanted to introduce precision, *rigorous calculations into rather unstable relations and above all on absolutely ideal bases. Most often, concrete observation played no role* and, when it did, it was only to lean in one direction: that of production of the cheapest deal, of the best equilibrium of purchases and sales. "Laissez-faire" was only limited by free competition and the two principles appeared in the eyes of liberal economists as moderating one another, thus resulting in *a compulsory adaptation of private interest to the general interest.*[21]

But on one point, the two principles, instead of leading to this dream equilibrium, accumulated their effects, became rivals, and produced fascism. Here is how this happened. If this equilibrium was working in theory, *the manufacturers sought by way of free competition to distort the equilibrium to their profit. However, due to "laissez-faire," they did not try this in the economy*, these doors being closed to them. But the economists hadn't foreseen that the practitioners, the manufacturers, found another means to open these doors: *technique*. Technique began to be developed alongside of and outside of scientific economy. Caught up in itself as it

---

21. In an economic context, the French *laissez-faire* roughly means "hands-off," implying that governments should abstain from policy intervention and should let markets find their own equilibrium of supply and demand.

was, this economy still neglected the enormous growth of production resulting from mechanization, or at least delighted in it, not seeing the danger to which this development exposed its very structure. This structure was built for certain quantities and for a certain production. Now these quantities are multiplied tenfold and production goes beyond the limits determined in the economic world. When the economists became aware of this, only two solutions were left: either totally rebuild the economic and social world (but this could no longer be achieved by theorists alone) or attempt a patch-up job and rein in technique to allow the liberal game to continue, i.e., the tangle of laws and the study of economic actions and reactions in a vacuum. This latter solution led us to *the fight against technique*. Nevertheless, this fight betrayed an upheaval in the old liberal economy, since the obsession was no longer to produce but to maintain prices high enough to allow businesses to turn in a profit.

This limitation of technical means happened in two ways: at first, in a disorderly fashion due to the inability of manufacturers to sell their products, an inability that was sometimes sanctioned by government purchases and the systematic singling out of products. On the other hand, and rapidly, it happened by more precise means: government orders, the diverting of techniques towards useless products. As for government orders, this was the system *of bonuses for non-production*, for the non-growth of factories, for the destruction of machines and their non-replacement, for the restriction of the productivity of new machines. These are all solutions that we can find in abundance in more or less all countries.

The only question seemed to be of maintaining high prices, even if the State had to pay to maintain them. I will speak as well of *the orientation of techniques towards sterile expenditures*. For there was a question which seemed incidental but soon became the main one: technique and rationalization entailed the discharge of a large number of workers. The liberals did explain that this was only a crisis to be followed by readjustment. Readjustment by the creation of new factories, by the creation of alternative work, which is to say that industrial workers diminished in number, and auxiliary workers grew in the same proportion. But precisely this growth in numbers was nothing other than the endless creation of useless work: advertising services, distribution services. Needed only because of competition and the saturation of products of technique, the market required an adaptation that was ever more difficult to achieve, caught between high production and high prices, an adaptation which itself entailed the employment of a constantly growing number of individuals to obtain it. In the final analysis, and this is the key point of the extreme development of the liberal economy, jobs and unproductive expenses multiplied. However, this multiplication did not manage to employ all discharged laborers. Unemployment resulted. After the abandonment of recent techniques, the abandonment of human workforces.

Having reached this point, the liberal era, to attain some order, could only call for a stable organization which would check these consequences as much as possible. *Command economy, organized capitalism*, whatever it is called, it is still, at a given moment, the very restriction that liberals allowed for from the beginning, of free competition and of laissez-faire, in the name of order and the common good. Besides, this organized capitalism did not only belong to liberalism in theory, but also in fact to the whole American liberal idea. The point was to build a new rational order of production, eliminating the misfortunes, the destitution, and the discord of the old capitalism, without abandoning the old liberalism, but by introducing new elements into it. Now the epitome of this modern liberalism was Fordism, the ideal circle of raising salaries by raising profits and raising profits by raising salaries. The eternal faith in this freedom which, once better reassembled and directed, must of necessity realize a brave new world.[22]

The play of liberalism was not interrupted in capitalism organized on these bases. And we have effectively stayed in the line of national liberalism, foreseen by the great liberals, a restricted and controlled economy. The facts are known. In all countries, production controls, distribution bodies, etc., have multiplied. This abandonment of the immediate principles of liberalism was accepted by liberals, since it was a matter of a readjustment. *The great principles, faith in the automatic arrangement of a mechanical world, were not abandoned.* But through this stage of command economy (a stage inevitably brought about by liberalism, as we have seen), we see fascism itself appear. For it is now enough that *these measures be stabilized, that the rules complement each other and become coordinated*. Soon we see an edifice that the State controls more or less completely, and which seems to be a fascist State. And so, we see here one of the first aspects of this continuity: *the influence of technique on the liberal economy.*

The second aspect, narrower and more specific, is no less important since it also leads to the fascist political state. It is fitting firstly to posit that

22. This last phrase in French is *le meilleur des mondes*, literally "the best of all worlds," the title of the French translation of Aldous Huxley's dystopian novel *Brave New World*. This novel left a lasting impact on Ellul, who referred to it frequently throughout his career. It was read by Ellul and Bernard Charbonneau when it came out in 1932 as confirmation of the latter's insights on technique, which Ellul would take up and eventually develop. Things came full circle when Huxley, recognizing this development of points he had tried to make in his novel, arranged for the English translation of Ellul's study *La Technique ou l'enjeu du siècle* (Paris: Armand Colin, 1954). See the statement from the publisher in the first English edition of Ellul, *Technological Society*, iii-iv.

the trade union is a necessary product of liberalism. By the simple play of economic forces, the labor union tends to form. *For liberalism assumes economic forces that are completely separated from their material contingencies.* When liberalism speaks of free agreement between the worker and the entrepreneur, it assumes equality between them. This having turned out to be false, to safeguard this equality, it was necessary to allow the weak to unite against the strong, so that the trade union appeared as a necessity. But since we are in a regime of freedom, trade unions are very many and defend very specific interests, their decisions are only valid for their members. They are, as it were, within the liberal game as a more complete but unique piece, exactly like the owner and the entrepreneur. The equilibrium is supposed to take place on its own here again, between the entrepreneur and the trade union.

From this situation, a double process would develop, leading straight to fascist corporatism. This process comes first from the *will to power of the trade unions, then from the agreements between trade unions and owners*. For trade unions are looking for a bigger mass. This is obtained by the merger of small trade unions, but it is elementary to observe that the larger a trade union is, the less active it is. The reasons for this are many. I see two main ones: first of all, the more the trade union grows, the more it encompasses different interests which, being less and less precise, are less and less urgent, and are slower to drive it to act. On the other hand, the State takes the big trade unions into account, includes them in its calculations, binds them to itself with favors, and then a certain bourgeois wisdom leads the trade unions to grow weaker in their action. In any case, once the trade union reaches a certain size, it is nothing more than an element of the economy, but not the moving element that constantly reacted to this economic force as a counterweight. It has become a stable element, with predictable reactions, which subsequently can be used whenever the State, in its goal of economic readjustment, decides to use it.

The State will be able to use trade unions all the more easily as it will have bound them even more to its form. In addition to the trade unions' desire for power, the permanent agreements between trade unions and owners accelerate this evolution. The most perfect model of this was that of the *company unions* in America.[23] Due to the division and rationalization of work, trades multiplied, and unions were born in great numbers because they were trade-based. Then groups of trade unions were created, under the bosses' influence, which united all the trade unions of one

---

23. "*Company* unions" is in English in the original, here and throughout this argument.

area of production with groups of bosses as governing bodies. This was a recovery of direct control of the trade unions, in the guise of giving them more power and abolishing the class struggle. But this centralization obviously offered a framework that was ready-made for the demands of organized capitalism. *With the State intervening to replace the bosses in their work of controlling the trade unions, intervening all the more easily since the trade unions were more powerful, all that was left to do was to carry out a relatively easy substitution.* In most cases, this substitution was made with the consent of the bosses themselves and sometimes with that of the trade unions. The latter then withdrew into their trade and declared, as in Germany, that except for the areas of salary and the organization of work, they were ready to enter into collaboration with the State. They gave up on any influence on government policies and thought they would find an ease of their operation in State control (Leipart).[24]

In fact, it must not be obscured that the whole system of *corporatism*, once it is stripped of misleading words, is nothing other than trade unionism crystallized in favor of bosses and the State. It meant a much more rigorous control of the person of the worker by means of the bodies that were initially meant to give him more freedom. The Italian corporation is nothing other than these *company unions* that Roosevelt was fighting. But the terminology changes. The words "bosses" and "workers" are suppressed, replaced with "directors" and "executives" or "leaders" and "followers."[25] It was claimed that in doing so the opposition between classes was suppressed. In suppressing the freedom of profit to replace it with professional control, it was claimed that profit itself was suppressed. The whole thing was covered up by formulas—for instance, that, thanks to the corporation, all men have the same civic right in national life. Incidentally, at the same time as this hardening of the framework of the worker's life, we see the birth of a legalization of big monopolies and trusts under direct control of the State. It hardly needs to be insisted that this is the inevitable counterpart of this hardening.

In short, from the economic standpoint, we see two branches of liberalism lose their internal strength to grow to become the objects of an external force and give birth by an inevitable process to a different economy, a command economy within the framework of corporatism. This is none other than the fascist economy. It is not in theories that this evolution was born. Mussolini, before 1922, had not thought of anything specific on this topic.

24. Theodor Leipart (1867–1947) was a prominent German union leader who worked for a merger of major trade unions in partnership with the emerging dictatorships of the Nazis in 1933 and the Communists in 1946.

25. In Italy and Germany respectively.

What is worse still is that, in fact, the corporations correspond well to the old trade unions, and not at all to the corporations as described by the party's philosophers and instituted by the jurists. There was only one corporation instituted according to juridical rules: the show business corporation in 1934. The law on corporations only goes back to 1934. Fascism, here again, appears as a simple modification of a state of affairs resulting from liberalism, but which was already nothing more than liberalism itself.

On the political level, the process is identical. Liberalism thought that State intervention in politics should be reduced to a minimum. The State should intervene only to maintain theoretical liberties and to protect the individual within the limits of the common good, thanks to the army and the police. The only rule to follow resided in the notion of public order. But it was soon realized that this maximum of theoretical liberty led to a maximum of oppression of certain individuals by others. One was led, in the name of liberalism, to accept State intervention to restrict the oppression of some. And this was not anti-liberal. On the contrary, the interest of the individual called for protection of the individual.

But on this point, liberalism parted ways with individualism. For by renouncing the certainty of a pre-established economy that was to be realized by a maximum of freedom, liberals began to have social concerns. From the political point of view, this social liberalism is represented by social democracy. The latter manifests, in a state of mind which is peculiar to it, a mere development of liberal evolution. There is neither an abandonment of goals, nor an abandonment of liberalism, but the certainty that the State personifies general interests. *Instead of admitting that particular interests coincide with the general interest, it is admitted that the latter should be defended by the State in the interest of individuals and their freedom.* Additionally, the State will enter the play of private interests to make it more precise and more equal.

But this stage of social democracy is already a preparatory stage of fascism. In this social movement, the vague fear of Revolution plays as well. A reformism on economic and political bases was thus created. Everything that you grant to the workers will be so much less in revolutionary programs. Now, this fear of revolution which drove liberalism down the path of social democracy is indeed precisely a liberal fact. One postpones the Revolution less because it is going to change something than because it will destroy the equilibrium which one has struggled to obtain, and above all, because by taking sides it will substitute a bias for liberalism's craving for objectivity.

How did social democracy announce fascism? With two signs: by the creation of new middle classes and by the dominant role in organization which it attributes to the State. Social liberalism leads to the creation, alongside the petty bourgeoisie deriving from the revolution (civil servants and rentiers) of a class of rich and comfortable workers. This is the American ideal. It is also the Bernsteinian reformism which contradicted Marxist theory, not on the fact of the creation of the proletariat, but on the fact that this new proletariat, while retaining exactly the same characteristics as the old one, could no longer be revolutionary because it lacked the desire to be independent.[26] The old petty bourgeoisie was replaced by this new class made up exclusively of subordinates, dependent on the higher classes, but whose dependence is voluntary and accepted because it offers material advantages which it would not have otherwise.

This may well be, incidentally, the most serious failure that socialism could suffer. This class could not be directly reached by fascism, because it could hardly be reached by this particular ideal of fascism, suppression of freedom, for instance. But if fascism initially relies above all on the bourgeoisie, *it has the means to reach this new middle class by using, in its turn, the methods of social democracy.* The latter become a method of rule which comes to conquer the mass of well-off workers. Furthermore, and on the psychological plane, it is certain that most of the myths of fascism easily affect this class, which after all brings about by its very creation a certain harmful spirit—cheap adventure, anonymous social duties, absence of guiding thought, of action, etc. If it is thus not an active element of the fascist fact, it is at least a passive element, ready to submit.

On the other hand, I have said of social liberalism that it gave an enormous task of organization to the State. But this mission is imprecise. We went from a set of rules that were pretty clear—the State in the face of war, or the issuance of banknotes—to a forced extension, because the economy escaped man taken in isolation. The State will, on the one hand, be in charge of organizing as well as it can interests which are too complicated to adapt to each other of their own accord. On the other hand (and this is only the complement of what has just been said), it will take charge of developing all national resources (incidentally, this is more or less what the liberals were saying when they entrusted the State with functions that private citizens could not fulfill).

As a result, it is actually very difficult to discern a limit where fascism begins, where the State is organized. To coordinate production and develop national resources, the State will be led to take increasingly strict measures as the economic questions become more and more complex. The theory of

---

26. Eduard Bernstein (1850–1932), German political theorist and socialist politician active in Marxist debates. For Ellul's treatment of Bernstein, see Ellul, *Les successeurs de Marx*, chapter 2.

State regulation of private industry, collaboration of classes in the corporative State—this is only the normal conclusion of liberal premises, once monopolies and State intervention are allowed. But on the other hand, this State can only perform its task with the complete submission of individuals. Faith in the State is necessary due to the State's economic development. The State becomes the absolute goal because it is the complete distributor. The "State above classes"—in the non-fascist state, this is called the "fatherland above parties" (Doumergue).[27] Once it is well catalogued and well accepted, we could substitute for the notion of the State that of the Fatherland and let out a great cry, invoking Hegel![28]

It would be rather curious to study the simultaneous birth of the nation and of liberalism. The fact of the nation had been taken into consideration before, but it does not seem that it entered into reality, as the expression of a whole. I obviously do not mean that the idea of the Nation produced liberalism, but I find this phenomenon strange enough to draw attention to it. For the nation assumed on the one hand an economy complex enough to be closed upon itself, and on the other, a certain mystique of the national which represented a superior value, the only admissible civilization. Now, it is the complex economy which will entail liberal expansion by the desire to bring to the maximum the group's resources and to develop them for the group. On the other hand, the idea of superior civilization is also, appearances to the contrary, a liberal idea. One must not forget that liberal civilization assumes that evolution stops at a kind of optimum conducive to the development of production and of the individual. One need then only assume a withdrawal of the nation into itself and the legalization of this belief in the superiority of the nation over others.

But we must go further. I am going to enter into a realm of more general politics. For currently, whether a State is authoritarian and fascist or non-authoritarian and "liberal," I hold that this may have some abstract, juridical, etc. importance, to please the professors of public law.

---

27. Gaston Doumergue (1863–1937), who was France's only Protestant president from 1924 to 1931, came back to power as prime minister of a conservative national unity government on the heels of the February 1934 rightist riots and leftist counter-demonstrations (in which Ellul and Charbonneau took part), a crisis that heightened the class polarization of French politics, leading to the 1936 election of the Popular Front.

28. "Invoking" translates the French *invoquant*, which could imply a simple allusion, or more strongly, a calling upon, a prayer. Considering Hegel's still-powerful influence in contemporary continental philosophy (see, for example, the wild popularity of the works of Slovenian philosopher Slavoj Žižek), Ellul's early mention of the explanatory power of Hegel's work for the evolution of the west may be worthy of careful examination.

But in practice, in the circumstances that we have described, this has only relative importance. The centralizing liberal State and the fascist State are following the same technique. *The trends of pure politics are no longer relevant because brutal facts, from which we can try to distance ourselves without success, impose certain forms of economy, certain forms of propaganda, and even certain forms of life.* Principles change, to be sure, and the German civil code will be based on new principles, but Krupp will remain at the head of a steel cartel because the steel cartel alone allows cheap enough sales, and steel exports competing with foreign steel enable the maintenance of a favorable balance of trade.[29] This favorable balance of trade enables the maintenance of the mark's value, which allows rearmament and well-organized propaganda, etc.

I could multiply causal sequences of this kind. All governments are in the same position. Liberals had no choice but to recognize that, as economics now leads the way, principled politics follows. Fascism recognizes this as well but makes pronouncements and posits principles to justify this admission. Saint Louis could give Guyenne back to the English but we cannot give Cameroon back to the Germans because there are land grant companies, shipping companies, insurance companies, and railway companies that would have to be reimbursed or bought back.[30] Mussolini, for having neglected these technical rules a little too much, is on the brink of bankruptcy (barring the intervention of foreign money). And this equalization of all forms of government before these necessities seems rather typical of our era.

We can no longer imagine a government which would neglect the press. This means of myth creation is quite evidently the very means of government but it imposes its type of government, abstract, ideal and uncontrollable, without which its role would no longer be necessary. A government that would fail to employ it would be at the mercy of a campaign led from outside, precisely from that point where control and the concrete act of knowledge cease. We thus see the formulas of politics lose their meaning and turn into dead beetle carapaces, apparently intact but

---

29. Krupp was a family-owned German steel company founded in 1587 that became the biggest company in Europe in the nineteenth and twentieth centuries, largely as Germany's main weapons manufacturer. It had long been a household word for the German military-industrial complex by the time Ellul began writing. It would later turn into the ThyssenKrupp conglomerate after economic problems led it to merge with another German steel giant at the turn of this century.

30. Ellul is referring to the Treaty of Paris (1259) in which the French King Louis IX allowed the English King Henry III to rule over Guyenne, which happened to include Ellul's home city of Bordeaux. Cameroon, which had been a German colony from 1884 to 1916, was divided between Great Britain and France following World War I.

now holding substances of another nature, inert sand, minute, necessary, and innumerable technical achievements.

The Passage from Liberal to Fascist Society

Finally, we must see how the passage from liberal society to fascist society takes place. To be complete, this would entail a preliminary study of the private juridical rules of these societies. Then we would have to consider society as a group of people ruled by certain institutions and, finally, to follow the evolution of these institutions and discern the physiognomy of the group in the one case and the other. I will only deal with this last point, namely, the social group aspect.

On the juridical problem, I will only make two remarks. The first one has to do with property. It could be just a riddle. Here is a text:

> In the (corporate) State, private property is respected. There is no expropriation without indemnity. The State nevertheless retains the right to limit and direct the use of the means of production and to intervene in the distribution plan according to the public interest. It does not have the objective of seizing the means of production, any more than in any capitalist country. Private property must remain the rule, and State property, the exception. Individual initiative is not replaced by State intervention. But the State retains the right to substitute for private initiative each time that this is necessary, to prevent it from developing in directions which would be harmful to public interest, and to direct it to obtain maximum benefit for the whole community.[31]

What a beautiful liberal harmony reminiscent of Adam Smith—a community of interests between property-owners and the community, even an identification of both interests under the paternal eye of a State that is only responsible for a work of direction and substitution, the liberal meaning of property. And yet these lines are taken from *The Economic Foundations of Fascism* by Einzig. Not much has changed in the new system and the old juridical notions are perpetuated.

The same goes for contracts, this second column of liberal legalism. We now see the notion of the autonomy of the will, which was all the rage,

---

31. Einzig, *Economic Foundations of Fascism*. Einzig (1897–1973) wrote this book to explain (or, in Ellul's eyes, ideologically justify) fascist economic doctrine for English contexts. A 1934 review of the book described it as "a thoroughly discreditable piece of journalistic hackwork . . . worthless for the purpose of serious study of the subject with which it purports to deal." Marget, "Economic Foundations of Fascism," 828–30.

taken up by fascism. The contract is nothing but the expression of the individual within the framework of law. Now fascism too is, on a larger scale and more totally, nothing but this same expression of the individual within this framework of the State and of law. The will remains the creator of law based on laws, and all the old liberal theories can reappear on this base alone. That is why I do not give much credence to the juridical transformations of fascism. It may make mass special laws against the Jews and others, proclaim that it is only based on the morality peculiar to its people, or refuse equality as a foundation of law and prefer the force which is realized within certain limits to some amorphous justice, but in spite of all this, Fascist law changes nothing. Property and the autonomy of the will are much more powerful realities than these formulas, and we find them in this allegedly new law. I will not dwell on this.

The most difficult point is the study of the social forms of our two societies. We cannot study them in detail here. Let us consider the different connections which individuals maintain in and through a social whole, under the liberal form and under the fascist form.

Let us recall that two kinds of solidarity need to be distinguished (with Durkheim, I take the distinction under its most primitive form): a mechanical solidarity and an organic solidarity. In mechanical solidarity, the individual is coagulated to society, directly, with identical prejudices. He almost completely alienates his personality in favor of society. But the word "alienate" seems false because it is not by a voluntary act, *but by his very nature* that the individual thus finds himself subject to society. It is not a question of a choice but of a state. At this moment, the individual is, as it were, unaware of himself. He is part of a collective consciousness of which he is but a momentary expression.

The collective consciousness thus plays a double role in relation to the individual conscience. On the one hand, it plays the role of a kind of indisputable and transcendent natural law which completely dominates individuals, for whom it is a kind of untouchable truth. On the other hand, it totally integrates the individual consciences and leaves them no freedom of action or of judgment. They cannot judge except via the criteria of the collective consciousness. In short, this mechanical solidarity is characterized by three features: first of all, by the fact that the individual is tied to the society as a whole, directly and without intermediaries; then by the fact that all his materials and methods of thought are provided to him by the collective consciousness; and finally, by the fact that the more this collective grows, the more the personalities of the members of the group diminish.

The most complete expression of this in society is penal law, as an expression of repressive law in general. Repressive sanctions are, in this form of society, the means to protect social resemblances. If there is repression, it is because there is a crime, and this crime consists essentially in a rupture of the mechanical equilibrium, of mechanical solidarity. Thus the more powerful this mechanical solidarity, the more it will tend to impose only one type of individual. Members of the society resemble each other more and more, and this resemblance is imposed under penalty of sanctions, with repressive law dominating all other forms of law.

Opposite this mechanical solidarity, we have organic solidarity. In this latter, individuals retain their personality, and organic solidarity is even intensified by the increase in individual personality. It assumes a coming to consciousness of social necessity and a kind of voluntary act which would consist in the sacrifice of part of the person to society. But this assumes a much more nuanced, less general, and less abstract collective. Society can no longer be a whole, but it will be fractioned into numerous sub-groups in which individuals will find their good, their center.

On the other hand, with the individual thus keeping a kind of graduated autonomy, going from the complete autonomy of one part of his person to a complete adherence to society through adherence to beliefs of varying degrees of generality, one sees that the individual plays an important role because a considered adherence is constantly at stake. One can no longer speak of pre-existing collective beliefs. These are more or less conscious and are formed little by little as a creation of the will. They thus no longer appear as an abstract and superior element, but as immediate and concrete. To better oppose this second solidarity to the first form of our three terms: (1) the individual here is no longer directly tied to society as a whole but to its parts, (2) collective beliefs are differentiated according to the functions that need to be fulfilled. There is a parallel strengthening of the personality and of organic solidarity, and (3) finally, the model expression of this solidarity is to be found in restorative law, that is, the law which expresses an equilibrium between individuals taken as particular individuals, as in civic law, for example. The model par excellence of restorative law is the contract. This restorative law is incidentally expressed either in a law of reality, which will be a relation to things (property law), or in a law of cooperation (commercial, administrative law, etc.).

There is no necessary chronological succession from the first type to the second. Neither of the two is tied to a specific social type, for instance, the mechanical solidarity of an authoritarian society or the organic solidarity of a liberal society.

But then *there is no opposition between individualism* (or rather, personalism) *and universalism, as the fascisms would have it*. One cannot separate the individual from the whole social group and the evolution of this group. To have attempted this separation and to have wanted to consider the individual in himself, on the one hand, and the society in itself on the other, was tantamount, within an artificial ideology, to separating two necessary elements of a synthesis. One lost sight at that point of the existence of this synthesis, and these detached elements were viewed as though they had a life of their own, identical to their real life. It was supposed that the individual taken in his pure state was identical to the individual immersed in society and the laws that were derived by reasoning about this isolated individual were thought to rightfully apply to the social individual. The same error, only in reverse, was happening with society. A society without people, living by and for itself and which revealed laws of the social body without repercussions, it seemed, on the individuals composing it. When fascism proclaimed the superiority of the social body and of the State, it was merely asserting more forcefully this separation that liberalism has prepared in favor of the individual.

On the contrary, from the moment that the action and reaction of individuals and of the social body are viewed as being characteristic elements of sociability, the forms of sociability will depend on the modifications of these actions and reactions. Now, these modifications take place in time. Onto the division of mechanical vs. organic solidarity is now grafted a division of the social body according to duration, where the mass can be distinguished from the group and from the abstract collective (von Wiese), with the *mass* being an essentially temporary unity, the *group* having a certain duration based on the impression that the members of the group can have of feeling bound to each other in a very concrete fashion either by a duty or by a goal to attain—and *the abstract collective* being, for its part, permanent, and based on a notion of mission of an ideological (and thus abstract) order; a notion common to all the participants and which they consider eternal.[32] Of these three notions, only one is necessary for my purposes: the notion of the mass. I will thus retain, aside from the division between organic solidarity and mechanical solidarity, the notion of mass which combines with it, insofar as it could represent one just as well as the other.

In order for there to be a mass, three conditions have to be met: a group of people differing in condition, nature, etc., who entertain some

32. Leopold von Wiese (1876–1969) held the first sociology chair in Germany and headed the German Sociological Association several times over his career. He highlighted the social processes (hence the temporal dimension) of human relations and how social constructs form structures—an emphasis also found in Ellul's sociological oeuvre. It might therefore be worth revisiting Wiese's *Beziehungslehre* ("doctrine of relationships"), which fell into oblivion in post-war academic sociology, even though Wiese is considered the cofounder of "formal sociology" alongside the more famous Georg Simmel (1858–1918).

representation of unity (though this unity need not be of a long-term necessary character); to be distinguished, consequently, from the crowd, or from the horde. The meshing representation of the unity of all the individuals of the mass can have very different reasons: a common interest, an economic or social situation (e.g., a group of unemployed people), or a feeling caused by the outside world, either of satisfaction or of discontent (e.g., the crowds of February 6).

We thus perceive that it is necessary to distinguish between abstract masses and concrete masses. *Abstract masses* are those which passively receive external influences or suggestions, which are identical for all. They are but a mass whose expression resides outwardly, in statistics, and inwardly, in the reaction that an individual within this mass can have to a phenomenon, a reaction that happens to be identical to that of any individual in this mass (e.g. the viewers of a film, the readers of a newspaper [Gurvitch]).[33] Their mass is indeed abstract, because they have no idea of the identity of their reactions, their role consisting in no longer being anything but a receptor which will in turn emit certain stimuli. Their representations will never be more than a coming to consciousness of this mass, and not a break with it.

However, only this coming to consciousness would risk preventing the passage from abstract mass to *concrete mass*. For let us suppose that an individual experiences, in uninterrupted succession, the creation and destruction of participation in various masses (office, cinema, café, newspaper, jazz). We will see gradually taking place the production of a complete integration of the individual in these successive masses. A mechanical solidarity is born. Now suppose that such an individual receives a sufficiently strong stimulus within whatever given mass to proceed to exteriorization, and, for example, to action. Because he is in the same state as all the individuals who make up this very precise mass (like the reader of the daily news), all the individuals of this mass will respond identically to this stimulus. Even without an individual command, all the readers of *L'Action Française* will congregate at the Place de la Concorde on February 6.[34]

Let us proceed with our suppositions. If all individuals belong to identical masses which take up their life entirely, if, consequently, they live in a

33. Georges Gurvitch (1894–1965) was a Russian-born French specialist of the sociology of knowledge and the sociology of law, of which he distinguished a wide variety according to types of social interaction, also stressing social rights as a complement to individual rights.

34. *L'Action Française* was a right-wing nationalist political movement aimed at restoring traditional French monarchy. On February 6, 1934, there was a protest against parliament at Place de la Concorde in Paris in reaction to the dismissal of a police prefect in connection with a financial fraud and Ponzi scheme in which several public figures and politicians of the Third Republic were involved.

state of abstract mechanical solidarity, and if these individuals receive the necessary stimulus, they all react in the same direction. But here, this will no longer be just the behavior of one evening, this will be a global exteriorization in their life itself. They will become the expression, no longer of a series of abstract masses, but of a series of actual, realized, concrete masses, which is exactly what is called fascism.

And so we find ourselves back in the heart of our question: *fascism appears, from the standpoint of forms of sociability, as a transformation of abstract masses into concrete masses within a mechanical solidarity.* But this is, after all, the synthesis of all that I have said up to this point: *liberalism and individualism prepare this transformation by a creation of abstract masses and by a constantly growing mechanical solidarity.* For it may well be said that all liberals were wrong in thinking that their doctrine led to increased individual self-awareness. Instead of seeing man, they saw stick-figure outlines of man, and their doctrines were based on these stick-figure outlines. The break in the frameworks that was attempted in the name of these outlines has only yielded a nearly pathological void. All the systems that the recognition of a sovereignty should have allowed to operate have disappeared. And the constitutional bodies that are collapsing due to liberal individualism now yield a brutal opposition between the individual and the sole sovereignty of the State. The sub-groups still resisting, like the family, are atrophied. Now, by virtue of liberalism, we are really in the presence of one of the features of a mechanical society. Instead of finding ourselves in a *civitas solis*, we find ourselves before a model of society which Durkheim assumed could only be found among Australians and Fuegians.[35]

We have seen that the most developed, extreme function of a mechanical society is the repressive function. Now what is developing, instead of the repressive function, is a preventative function. In the name of common sense, in the name of the common good, in the name of common morality, a type of common man (*homo rationalis vulgaris*, as the *Petit Larousse* will one day call it) tends to be created.[36] And we tend toward this both

---

35. *Civitas Solis*, Latin for "City of the Sun," refers to a 1602 utopian work of that title by the Dominican friar Tommaso Campanella (1568–1639). Émile Durkheim (1858–1917) was one of the founders of modern social science. In 1887 he was appointed to the faculty of the University of Bordeaux and taught there the first course on social science. In 1895 he established the first European department of sociology at Bordeaux and published his classic manifesto, *The Rules of Sociological Method*. In 1902 he was appointed professor at the Sorbonne in Paris. His 1912 book *The Elementary Forms of Religious Life* bears the subtitle "The Totemic System in Australia."

36. The *Petit Larousse illustré* is a compact encyclopaedic dictionary that was, for many decades after its first edition in 1905, the go-to reference work in every French-speaking household, classroom, and workplace.

voluntarily and involuntarily: voluntarily by the creation of an internationalism of morality, of law, of civilization, of literature, of art which gives averages of morality, juridical statistics to determine the value of a legal rule, *a priori*-ism in civilization, all-purpose literature and utilitarian surrealism in art. These notions are either too abstract or too elementary, obedient both to the oft-denounced scission between the real and thought, and to the belief in the superiority of large numbers. But they are pernicious notions in any case, because they become implanted in people's lives, and those who do not accept them are considered abnormal. The abnormal are to be identified and cared for in observation facilities, houses of correction, etc. A personal domain is left for them so long as this personal domain in no way changes their character, nor their life, but is limited to dreams, to the ideal, to the "mystical," to introspection, etc., to everything directed towards the inside, and never goes out. Once this ideal man is created, its model will be spread on every occasion by the enormous means of persuasion at our disposal. A million men cannot be wrong, declares a shaving cream whose name I forget. They are accepted truths. Opening a newspaper is enough to breathe this air, a women's advice column or matrimonial personal ads.

Liberalism has brought about a social amorphism which is probably without historical precedent. It has allowed the creation of these abstract masses of which I have just spoken, of this life in masses and solely in masses, where the life of man is wrapped within a series of overlapping circles which totally absorb the individual. Café group and club group, sports group and occupational group. He takes on one character in one place, and another character in another milieu. He is no longer himself, he is essentially the social man, obtained by preventative means, the one from whom society no longer has anything to fear; on the contrary—he can only stabilize it, and this is certainly what will happen.

In this neo-mechanical society, the shock that will bring about the appearance of concrete masses will be all the easier as the amorphism becomes more complete. And likewise, the notions of sacrifice and heroism will be all the more easily exalted as the individual loses a sense of his own value. Fascism appears, from the social standpoint, as a better-arranged amorphism, more voluntary than the other, liberal one, but of the same nature, belonging to the same type of society. This type could be called inferior, even though it is rather difficult to pass a value judgement on social forms. It is a type of mass which will never attain the superior type of the group, since there is no genuine organic solidarity in this inchoate society; and still less that of the abstract collective, which assumes for everybody a personal and individual coming to consciousness of the mission that the group may have to fulfill. Of course, this can seem paradoxical after what I have said of the mass, which is an essentially temporary gathering, but we must not forget that the technical means at the disposal of governments allow them to keep crowds in this state of artificial and temporary union, as in a kind of icebox. There is a constant recreation of the mass by exterior means.

Fascism is thus the worthy son of liberalism. It keeps all the features of its father, but along with the features of its mother, technique. Liberal society was rolling along on its own and seemed balanced, when an element came up to trouble this balance, namely, technique. Liberal society was not made for it. It reacted, trying at first to integrate, then to stop technique. It could not assimilate it because technique was in full progression, while society seemed fixed within bounds which it liked to think were unalterable. The adjustments liberal society made to technique broke apart one after another. It then tried to master them, but for this, it had to use violent means, and above all it had to know the point to which things had come.

Liberal society was incapable of this self-evaluation because its methods and its wishes were still too vague and uncoordinated. In the face of this need to take stock, in the face of the prevailing common mentality, only one reaction was called for: fascism. Of course, the development of technique was merely its material condition, and we have already considered all the intellectual factors and all the spiritual resignations which were necessary to come to this point, and which liberalism had long been preparing. That is why, to play its role, fascism will have to be essentially demagogic, which, in fact, it is. It will have to take up and proclaim all the commonplaces of liberalism, reassemble all the juridical and intellectual creations of liberalism and raise them to the level of institutions. It will have to present itself as the factor of reconciliation of all average individuals around an average individual who will represent them better than each of them. It is essentially a stabilizing element which, having found a situation, takes up all the elements of this situation, puts them in order, reclassifies them. It petrifies what was struggling and gives an exact account of what is done. It is a pause within a decadence. It is a period of disappearance of liberalism, but not at all because it has reacted against liberalism: *because it arose in the decadence of liberalism to confirm this decadence and bring it to the knowledge of all.* We cannot see in fascism a mere suspension of the normal evolution of liberalism towards another form, perhaps wholly different. This ring which fascism has forged around the decaying values of liberalism is not made to be eternal. If we can have confidence in some power, it is in technique itself which has been the occasion of this ring, but which will also be the factor of its breaking. Technique has neither reasons nor any capability to stop. To regain control over it, man would have to take a distance from it himself, and it is not fascism that will help him there. As for us, for now, we must watch it accomplish its work, and wait, for we are too small.

De Tocqueville gave an admirable description of fascism when he wrote roughly one hundred years ago: "

> Democratic societies which are not free may be rich, refined, ornate, even magnificent, and powerful in proportion to the weight of their homogeneous mass. They may develop private virtues, produce good family-men, honest merchants, respectable landowners . . . but there are things which such societies . . . can never produce, and these are great citizens, and, above all, a great people . . . (because) one's love for despotism is in exact proportion to one's contempt for one's country.[37]

Bibliography

De Rougemont, Denis. *Journal d'une époque 1926–1946*. Paris: Gallimard, 1968.

De Tocqueville, Alexis. *The Old Regime and the Revolution*. Translated by John Bonner. New York, NY: Harper & Brothers, 1856.

———. *The Old Regime and the Revolution*. Vol. 1, *The Complete Text*. Edited by François Furet and Françoise Mélonio. Chicago: University of Chicago Press, 2001.

Durkheim, Emile. *The Elementary Forms of Religious Life*. Translated by Carol Cosman. Oxford: Oxford University Press, 2008.

Einzig, Paul. *The Economic Foundations of Fascism*. New York: Macmillan, 1933.

Ellul, Jacques. *Autopsy of Revolution*. Translated by Patricia Wolf. New York: Alfred A. Knopf, 1969.

———. "Le fascisme, fils du libéralisme." In *Les années personnalistes*, edited by Patrick Troude-Chastenet, 113–37. Cahiers Jacques Ellul. no. 1. Bordeaux, 2003.

———. *In Season and Out of Season: An Introduction to the Thought of Jacques Ellul*. Translated by Lani K. Niles. San Francisco: Harper & Row, 1982.

———. "Le Personnalisme et Mounier: Pourquoi je me suis séparé de Mounier." *Réforme* 265 (April 1950) 6–7.

———. "Le Personnalisme, révolution immédiate." In *Les années personnalistes*, edited by Patrick Troude-Chastenet, 81–94. Cahiers Jacques Ellul. no. 1. Bordeaux, 2003.

———. *Les Successeurs de Marx: Cours professé à l'institut d'études politiques de Bordeaux*. Edited by Michel Hourcade, Jean-Pierre Jézéquel and Gérard Paul. Paris: La Table Ronde, 2007.

———. *The Technological Society*. Translated by John Wilkinson. New York: Vintage, 1964.

———. *The Theological Foundation of Law*. Translated by Marguerite Wieser. New York: Doubleday, 1960.

---

37. de Tocqueville, *Old Regime*, x–xi. (A more recent English text of *L'Ancien Régime et la Révolution* (1856) is available as *Old Regime*, vol. 1, *The Complete Text*.

Ellul, Jacques, and Patrick Chastenet. *À contre-courant: Entretiens*. Paris: La Table Ronde, 2014.

———. *Jacques Ellul on Politics, Technology, and Christianity: Conversations with Patrick Chastenet*. Translated by Joan Mendès France. Eugene, OR: Wipf & Stock, 2005.

Greenman, Jeffrey P., Read Mercer Schuchardt, and Noah J. Toly. *Understanding Jacques Ellul*. Eugene, OR: Cascade, 2012.

Loubet del Bayle, Jean Louis. "Aux origines de la pensée de Jacques Ellul? Technique et Société dans la réflexion des mouvements personnalistes des années 30." In *Cahiers Jacques Ellul: Pour une critique de la société technicienne no. 1, Les années personnalistes*. Bordeaux: Pixagram, 2004.

Marget, Arthur W. "The Economic Foundations of Fascism: Paul Einzig." *Journal of Political Economy* 42 (December 1934) 828–30.

Menninger, David C. "Marx in the Social Thought of Jacques Ellul." In *Jacques Ellul: Interpretive Essays*, edited by Clifford Christians and Jay Van Hook. Chicago: University of Illinois Press, 1981.

Rognon, Frédéric. *Générations Ellul: soixante héritiers de la pensée de Jacques Ellul*. Geneva: Labor et Fides, 2012.

———. *Jacques Ellul: Une pensée en dialogue*. Geneva: Labor et Fides, 2013.

Roy, Christian. "Aux sources de l'écologie politique: Le personnalisme gascon de Bernard Charbonneau et Jacques Ellul." *Canadian Journal of History* 27 (April 1992) 67–100.

———. "Ecological Personalism: The Bordeaux School of Bernard Charbonneau and Jacques Ellul." *Ethical Perspectives* 6 (April 1999) 33–44.

Spengler, Oswald. *The Decline of the West: Form and Actuality*. London: Forgotten Books, 2017.

Villari, Luigi. *The Fascist Experiment*. London: Faber & Gwyer, 1926.

Virilio, Paul. *The Administration of Fear*. Los Angeles: Semiotext(e), 2012.

2

Bernard Charbonneau, the State, and Politics

BY Daniel Cérézuelle

> *Daniel Cérézuelle was born in Bordeaux in 1948. He studied the philosophy of technology with Jacques Ellul (Bordeaux), Hans Jonas (New York), and Jean Brun (Dijon). He also had many friendly and intellectual exchanges with Bernard Charbonneau. His 1979 Dijon doctoral dissertation studied* Le mythe de la technique: Généalogie des idéologies technicistes. *He has taught philosophy of technology in France, the United States, and China. He is Scientific Director of the Programme Autoproduction et Développement Social (PADES) and serves on the board of the Société pour la Philosophie de la Technique. Among his books are* Ecologie et liberté, Bernard Charbonneau, précurseur de l'écologie politique *(2006) and* La technique et la chair; essais de philosophie de la technique *(2011). As a sociologist he is investigating the social importance of the non-monetary economy in a modern society.*

* * * *

"UNITED BY A COMMON thought." These are the words of Charbonneau (1910–1996) when, after Ellul's death in 1994, he wrote an account of his life-long friendship with Jacques Ellul.[1] Both held as vital the values of freedom, autonomy, responsible action, and incarnation of truth in reality. As early as the 1930s they shared a strong intuition that modern industrial civilization, with its huge cities and industrial productive equipment managed by state bureaucracies, excludes the possibility of living according to these values. Instead, industry, technology, and

1. Charbonneau, "Unis par une pensée commune."

bureaucratic management of society foster a depersonalization of daily life. Both were convinced that this trend should be resisted.

> I still can visualize one night when we were pacing from his doorsteps to Fondaudège street and back, passionately discussing the meaning of our lives in a world drifting towards the worst, and the necessity to change it. Freedom, for him as a Christian, and for me simply as a human being, seemed to be imperiled on two levels—in the short term by the rise of totalitarianism, from the left or the right, and in the long term by the means of propaganda and repression which technology made available to the corporations and the state. But at a time when the threat of war and revolution dominated political passions, except for a few individuals, such a question was unthinkable.[2]

Charbonneau was twenty-five and Ellul twenty-three in 1935 when they wrote their "Directions for a Personnalist Manifesto," in which they propounded ideas which they would hold and develop throughout their lives.[3] First, they argue that the dominant force in our world is technology (*la technique*). It is technological progress which gives rise to the urban, industrial, capitalistic, and institutional concentration characterizing modern life. By "technology" they do not mean only machines but a "general way" (*procédé général*) of addressing all human problems (this would be developed in Ellul's 1954 book on technology).[4] From the unchecked development of modern technique/technology result new social constraints, among them bureaucracy and the expansion of the state. Against these developments, politics, understood as control of government and of the state apparatus, is powerless. This is an idea which Ellul later developed in his 1965 work on politics.[5]

Therefore, a revolution is necessary in order to change the structures of society, especially technology and the state. Such a revolution should not focus on the rejection of this or that ideology, such as communism or fascism. It should be a revolution against the trend towards the totalitarian, all-encompassing state. "Salvation will result from the diminishing of the effective power of the state, not from its reinforcement or from the suppression of its apparent power, nor from the creation of a super-state."[6] A real

2. Charbonneau, "Unis par une pensée commune," 19.
3. Charbonneau and Ellul, "Directives pour un manifeste personnaliste."
4. Ellul, *La Technique, ou l'enjeu du siècle*, published in English as *The Technological Society*.
5. Ellul, *L'Illusion politique*, published in English as *The Political Illusion*.
6. Charbonneau and Ellul, "Directives pour un manifeste personnaliste," §57.

change in our civilization requires that we rebuild society from the bottom up. This is the reason why they advocated regionalism and federalism, the control of technology and of the power of money.

In order to reduce the power of the state and technology in our social and personnal life, we must first understand the logic of their development as well as the origin of their dynamism. A critical theory is necessary. Both Charbonneau and Ellul were convinced that having an intuition of the nature of the obstacle (the state, technology, propaganda, etc.) is not enough. We must also understand how and why these obstacles are so powerful.

In personal conversation, Ellul told me that in 1945, after the war, he had a discussion with Charbonneau and they decided that there were two main issues which should be dealt with in depth: technology and the state. In fact, it was Charbonneau who decided that Ellul should focus on technology and that he would take on the task of writing about the state. "Bernard had a strong influence on me," Ellul told me, so he complied. In fact, Charbonneau had already begun to work on the issue of the state during the war and his book, entitled *L'Etat*, was completed by 1949.[7] This book, however, was rejected by publishers in spite of Ellul's efforts, and was only published in 1987, almost forty years later. Ellul thoroughly agreed with Charbonneau's analysis of the state and of politics. Nevertheless, in his own *Political Illusion*, his three books on revolution, and his monumental *Histoire des institutions*, Ellul uses very different language and ways of reasoning.[8]

Charbonneau's Main Ideas on the State and Politics

The prospect of a social totalization. In *l'Etat*, Charbonneau describes how the technological and industrial dynamism of liberal society has created the conditions for a total technocratic organization of collective and individual life. This book, written in the context of the World War Two, was completed two years after the collapse of the Nazi regime and at the time of the triumph of Stalinism. By analyzing the slow development of the state in western civilization, culminating with the contemporary political horror, Charbonneau aims to warn us against an unthinkable future, the progressive establishment of a total social (not simply political) order, of which the totalitarian regimes of the twentieth century are an example.[9] "Through its crisis, the totalitarian

---

7. Charbonneau, *L'Etat*.
8. See bibliography for publication data.
9. Many years later Charbonneau discovered that Bertrand de Jouvenel had developed a similar analysis in his great book *Du pouvoir, histoire naturelle de sa croissance*; ET: *On Power, Its Nature and the History of Its Growth*.

state evolves towards a total order where the constraint of the state would become unnecessary."[10] Where most authors, such as Hannah Arendt, try to explain "how could we get into this situation?," Charbonneau asks "what does this horror preclude?" What is important for him is not so much the totalitarian ideologies which justify limitless political power, as the totalitarian *phenomenon*, the real possibility of a social totalization.

Political ideologies are secondary causes. Since the Thirties, Charbonneau and Ellul held a very original position, arguing that political ideologies are not the decisive factor of political evolution. We miss its essence by by focusing on political ideas. They were convinced that it is first at the level of daily attitudes, of ordinary social behavior and beliefs, of commonplace ideas which are born in liberal society, that the totalizing role of the state is made possible. This is why, when a political crisis occurs, liberal society is already prepared for a totalitarian regime.[11] There is nothing that totalitarian regimes have done which has not been done previously by liberal societies. Besides, the totalitarian phenomenon has been prepared at a deeper level by the slow development of social infrastructures: propaganda, industrial organizations, and various government and corporate bureaucracies. The growth of these infrastructures is largely autonomous, as exemplified by the history of the state.

According to Charbonneau, the state is not a neutral mediator or a simple tool for living together. It has its own logic of development and functioning characterized by autonomy, self-growth, centralization, all of which characterize also the development of modern technology. The state can be considered as a technique for managing human ensembles in an impersonal way.

> The state and the machine are just two different manifestations of the same kind of development. Through their unifying agenda, industry and the state converge on the same goal. Today they are on the verge of merging. In modern warfare, the firepower is industrial power. Economic concentration resulting from the progress of mechanization requires, sooner or later, political centralization. The power of big business precedes the power of the state. The same deep reason drives their progress: a will to material power.[12]

As a result, every state tends to transform the citizen into a replaceable element of an impersonal machinery. Soon the state grows "by itself." Its

---

10. Charbonneau, *L'Etat*, 414.

11. See "Fascism, Son of Liberalism" by Jacques Ellul in chapter 1 (above).

12 Charbonneau, *L'Etat*, 108.

operations require the reduction of individual responsibility. In order to be efficient, it must extend its control over society through law, regulation, administration, finance, and police. Therefore, to speak of a "liberal state" is a paradox. Pluralism, freedom, and creative, responsible action are not in its nature. Unavoidably, the development of the state exhausts the capacity of self-organization of society and local groups. And this social anemia calls for more progress of the state and its administration which tend to pursue their own goals. Regardless of political programs and social norms and values, bureaucracies develop their own specialized culture and interests, which reinforce the tendency of the state towards autonomy (the functioning of a modern army is a good example).

The progress of non-political structures fosters the growth of the state. The development of industrial economy, of technology, and of science, prepares for, and requires, the management of society by the state. What liberals call "progress" is the condition of a control by the state of the totality of social life in the name of efficiency. If we want more economic, technological, or military efficiency, then we need a powerful state. "The totalitarian regime could be defined as a sudden actualization of the social potentialities of modern technology."[13] More powerful tools require equally powerful and strict management; otherwise we risk social, environmental or political chaos. Besides, since technology is constantly transforming society it becomes necessary to adapt individuals and groups to an everchanging context. Equally necessary is the control of opinions and attitudes by propaganda, whose tools and techniques are developed in liberal societies. A technological society is by necessity a society of control. Charbonneau's explanation is worth quoting at length:

> Since all organizational progress is surrounded by a fringe of disorganization, like living flesh rotting at the contact of steel, organization becomes even more necessary. When it reaches a certain point, it disrupts the natural balance and grows by itself, either because it stimulates irreducible obstacles that it must destroy, or because it fosters passivity and therefore creates a vacuum that must be filled. The last spontaneous activities that concentrate all the potential for freedom turn into factors of disorganization. For example, when freedom exists only in leisure, then leisure time becomes so important that it becomes necessary to give it a formal status.
>
> Once organization has destroyed freedom, this destruction calls for more organization. The habit of receiving orders from above destroys the individual's sense of initiative and

13. Charbonneau, *L'Etat*, 350.

self-discipline, obliging authorities to interfere in new and unexpected areas. Then the process of organization gains momentum, tending to encompass everything

Perhaps we have reached the point where the technological enterprise can no longer stop halfway. It is condemned to rebuild artificially the natural totality broken by the endeavors of human freedom. When human power expands to impact the whole earth, it becomes necessary, for the sake of survival, that science penetrate the multitude of causes and effects that constitute the world. Technology and the state implement these conclusions with the power and the reach henceforth seen as maintaining creation. When we have gained the power of acting upon individuals and society, technology must not only replace ancient tools but also family, society, and God himself. "Science will organize society, and after having organized society it will organize God."

Beyond a certain level of organization, we can only choose between chaos and the system which reconstructs from outside the world that has been destroyed. All our uncertainties lead toward this total immobility. Then a technology of organization develops that transcends particular techniques: the technology of the totalitarian state. It results more from our reason than our passion. This total organization, claiming to realize the absolute through finite things, is the exact definition of antifreedom. Nevertheless, organization *per se*, is legitimate and the very thought which criticizes it is the offspring of the same spirit which pushed us to conquer the cosmos. Organization is legitimate because it cannot be separated from human freedom. We must organize to obtain a free life. Organization becomes illegitimate only when it tends to become total. It becomes total when we refuse to consider the contradiction, when we pretend that freedom is anti-freedom. At bottom, evil is less in total organization than in the total lie which justifies it."[14]

Dining With the Devil: How Much State Do We Need?

Charbonneau and Ellul shared the same values and had a common understanding of the goals of collective action and of the steps towards the "necessary revolution." They advocated for a social change induced "bottom

14. Charbonneau, *Le système et le chaos*. 49–50. Charbonneau is quoting Ernest Renan's 1890 *L'avenir de la science*: "Science will organize society, and after having organized society it will organize God."

up" through the action of a host of small local groups, developing their own institutions and thus creating a kind of dissident, counter-culture society within present society. Most of all, these groups should be intent on creating a different way of life (*style de vie*) which could be truly revolutionary and would render the state and its impersonal institutions less necessary. Nevertheless, in spite of the fact that throughout their lives they engaged in common actions, there is some difference in the way Charbonneau and Ellul considered the means of action suitable for implementing this common program.

Ellul and the "no dining with the Devil" option. Beginning with his earliest writings, such as *Presence in the Modern World*, Ellul stuck to the principle that we cannot achieve good ends with bad means.[15] Too often we are seduced by the search of efficiency and power into forgetting that bad means corrupt action, and betray the requirements of "incarnation." Therefore, in order to change society we cannot and should not use the state and its impersonal machinery, or political parties. This attitude was reinforced by Ellul's brief political experience as member of the Délégation municipale of Bordeaux in 1946.

In 1939, Ellul wrote to Charbonneau, who wanted to initiate collective action: "If we have to rebuild the world with the same elements, they are so independant from us that I am afraid they will overcome us, whatever our strategy might be."[16] There is an old saying that "He who sups with the devil should use a long spoon." But for Ellul it is better not to sup at all with the devil. In 1950, in another letter to Charbonneau, he raises again the same issue: "You know well that when I speak of compromising, I am not judging you, but I am afraid of the powers of this world, which corrupt from within. This is why as far as I am concernend, I keep away from them as much as I can, not because of asceticism, or rigorism, but because I do not trust myself, because I am weak."[17] In a 1951 letter, discussing whether Charbonneau should borrow money from a public bank, he writes "compromising with the state is never a desirable option."[18] Hence the politics of Ellul which, in the Proudhonian tradition, favor anarchist and federalist options, rely mostly on the spontaneous creativity of the people and are very defiant of institutionalization.

Ellul was a radical Barthian who believed that we live in a corrupt world where nothing divine remains. It is only when grace enables us to

---

15. Ellul, *Presence in the Modern World*, 39–62.
16. Ellul, Letter to Bernard Charbonneau. 1939.
17. Ellul, Letter to Charbonneau. 1950.
18. Ellul, Letter to Charbonneau. 1951.

follow the Word of God that our action may have a positive outcome. Unflinching on this principle, he wrote to Charbonneau in 1945, "I do not see any reason for a man to change his life. There is nothing inside and outside of him, which is not corrupt, not absorbed by the mass. I see that only God can act. Faith alone, because it is not from man, can convince man to resist death." (by "death" Ellul means spiritual destruction). Since Ellul knows that Charbonneau does not share his faith, he writes, "for me you remain an astonishing act of God."[19]

Charbonneau and the "long spoon" option. Charbonneau was always more pragmatic, hence his discussions with Ellul about compromising or not. For him, since we have to sup with the Devil anyway, we really need a long spoon. The question is "how long of a spoon?"

For Charbonneau it is clear that today the state is an enemy. This is clearly stated in *The Green Light*, again worth a long quotation:

> The same goes for the state: to free ourselves from it, we must first acknowledge it. On this point, the ecological left is divided between the supporters of "everything is political," borrowed from left-wing parties and the opponents of parties and the state. Christian or post-Christian ecologists take up the Marxist explanation that brings back all the sorrows of the earth to this *diabolus ex machina*: capital identified with property. Hence the solution: socialization. Unfortunately, in the East, socialism has only accelerated the ruin of nature at the expense of freedom, while in the West, the technocratic state, with expropriation at its disposal, is the ally and promoter of capital's designs. The "socialization of nature" advocated by ecologist Philippe Saint-Marc, former president of the Mission interministérielle pour l'aménagement de la côte aquitaine (MIACA), is a contradiction in terms. Society is the antithesis of nature, and the latter has nothing to do with management.
>
> The case of the development of the Aquitaine coast shows that it is even more efficiently destroyed by the coalition of senior civil servants and urbanists than by the promoters they attract in order to realize grandiose plans, inspired by the need to become famous as much as by greed for profits. When it comes to the destruction of nature—nuclear plants, military camps, highways, land consolidation, etc.—even in the West, the state is always front and center. As for cultural differences, they are abolished by public administration, the school system, and barracks as much as by corporations.

---

19. Ellul, Letter to Charbonneau. 1945.

> Thus, the anarchistic strand is not wrong to mistrust the state. But I fear once more that it confuses the struggle for absolute freedom with the one against the absolute state. In the first case, one fights to obtain everything, while in the second case, one does it in order not to lose everything. And if the struggle against the State has been going on forever, its suppression is but a dream that until now has only led to its reinforcement in the name of a withering away that we are still waiting for. Anarchy has done little more than throw bombs and rule literature and songs. The governance of any society of a certain size, especially if it is equipped with technical means, takes place through a state, which is centralized, or at the very least federal. One might as well know it if one wants to "confine" the central power within proper limits. The issue is not replacing the state by generalized self-management, but preventing the advent of the total administration. This can only be done by awakening at the grassroots level people and societies that will resist its grip, and by defining the faith and the common institutions that can federate individuals and societies that happen to be different.[20]

This quotation shows clearly that for Charbonneau there is a necessity of the state. He is not against any and all political institutions. He is against a state with no limits and he thinks that we should work at reducing its role as much as possible. Hence his statements in *L'Etat*:[21]

> Fighting the state cannot be a principle of action. Living free is the principle. If I insist on anarchy, it is because today all the threats against freedom are the result from the increase of its power. In other times, concern for freedom might have me fight for establishing a political order.
>
> A society without State is a utopia, the same with a man without sin.
>
> The State is our weakness, not our glory. This is the only political truth.
>
> It is impossible to suppress the state, but it is also necessary to reduce its power as much as possible.

There is no general rule for determining the "length of the spoon" and how to reach an exact balance between those two contradictory necessities. All we can do is to cultivate an acute consciousness of this tension.

---

20. Charbonneau, *Le Feu vert*, 105–7. Quotation from Christian Roy's forthcoming translation, *The Green Light*.
21. Charbonneau, *L'Etat*, 436–37.

Reflecting on ecological action and political ecology, consistent with the "long spoon" option, Charbonneau thinks that at some point the ecological movement might have to consider action through a political party and existing institutions. In *The Green Light* he explains that a green party could be useful on the condition that its power should be limited by other ecological institutions, on the one hand, by a federation of local grassroot organizations and, on the other, by an independant Ecological Authority (in the manner of the Fabian Society for British socialism), which would define the goals and remind everyone of the values. But this should be done with extreme caution, especially by avoiding any brutal change and any *tabula rasa* policy such as have sterilized many modern revolutions. According to Charbonneau, if we consider responsibly the possibility of governing society by means of the state, it will be necessary to begin by a management of the existing situation in order to change it.

A common thought, two different spiritual tempers. Ellul was a "Barthian anarchist" with a deep distrust of wordly means. His writings remind us of directions and conditions of action but show little concern for the creation of specific revolutionary institutions. Like Bakunin, who he sometimes mentions, Ellul believes in the possibility of a revolutionary spirit and in the creative power of people, capable of inventing in a short time new ways of living together and interacting with the world. Ellul is also an historian, (specifically of institutions) and in *What I Believe* he expounds his anthropology. For him, man is an historical being, capable of creative breakthroughs, and of taking risks. His is an anthropology of movement, of change, of adventure: "I am always convinced that man can initiate something different from what looked unavoidable."[22] But since nature as well as human reason are fallen, human creativity requires guidance from a transcendant perspective, from a God who is "wholly other," thoroughly different, unexpected, and surprising.

Once, Charbonneau told me "I am more catholic than Ellul. I emphasize more than he does the need for stable institutions." By training Charbonneau was not only an historian but also a geographer. Calling for change, he was more interested than Ellul in the search for a new equilibrium, a new balance between freedom, nature, and society. Such a balance can be achieved only slowly. From his concern for the diversity of agriculture and cities he drew a keen attention to the slow emergence of spatial structures finely adapted to local conditions.

It is significant that in the Thirties, when Ellul was inspired by Karl Barth who insisted on God's transcendance, Charbonneau was also interested

---

22. Ellul, *A contre-courant*, 75.

in the writings of Emil Brunner who insisted on God's presence. According to Brunner's doctrine of analogy, nature and human reson are not absolutely fallen and treacherous. This agreed with Charbonneau's emphasis on human sensibility and our ability to relate to the world through our flesh. For Charbonneau, sensuous experience provides us with an essential knowledge of the world and he would often mention that in latin *sapere* means both to taste and to know. Sense experience is a basis on which our reason can elaborate a critique of modern depersonnalization.

To illustrate this diference of spiritual temper in Charbonneau and Ellul I shall rely on two images from their homes which I have often visited. I remember that in Ellul's dining room there was a wooden cross hanging on the wall. In Charbonneau's living room there was a fireplace where he would light a fire every evening. In this fireplace there was a cast iron fireback which had been fashioned according to his own design, with the motto "my reason is flame." Of course, by "reason" he did not mean the cold capacity of logical calculation but a passionate desire for holding together truth and reality, freedom and nature.

Bibliography

Charbonneau, Bernard. *L'Etat*. Paris: Economica, 1987.
———. *Le Feu vert*. Paris: Karthala, 1980. English translation: Roy, Christian. *The Green Light*. London: Bloomsbury, forthcoming.
———. *Le système et le chaos*. Paris: Economica, 1990.
———. "Unis par une pensée commune." *Foi et Vie* 93 (December 1994) 19–28.
Charbonneau, Bernard, and Jacques Ellul. "Directives pour un manifeste personnaliste." *Cahier Jacques Ellul* 1 (2003) 63–79.
———."Directives pour un manifeste personnaliste." In *Nous sommes révolutionnaires malgré nous*. Textes pionniers de l'écologie politique. Collection Anthropocène. Paris: Le Seuil, 2014. Also in *Les Annees personnalistes*. Edited by Patrick Troude-Chastenet, 63–79. Cahiers Jacques Ellul No. 1. Bordeaux, 2003.
de Jouvenel, Bertrand. *Du pouvoir, histoire naturelle de sa croissance*. Geneva: Constant Bourquin, 1947.
Ellul, Jacques. *À contre-courant: Entretiens avec Patrick Troude-Chastenet*. Paris: La Table Ronde, 2014. English translation: *Jacques Ellul on Politics, Technology, and Christianity: Conversations with Patrick Chastenet*. Edited by Joan Mendès France. Eugene, OR: Wipf & Stock, 2005.
———. *Autopsy of Revolution*. Translated by Patricia Wolf. New York: Knopf, 1971.
———. *Changer de Révolution*. Paris: Seuil, 1982.
———. *De la Révolution aux révoltes*. Paris: Calmann-Lévy, 1972.
———. *Histoire des Institutions*. Tomes 1–5. Paris: Presses universitaires France, multiple editions.
———. Letters to Bernard Charbonneau. Charbonneau papers, Institut d'Etudes Politiques de Bordeaux.

———. *The Political Illusion*. Translated by Konrad Kellen. New York: Knopf, 1967.
———. *Presence in the Modern World*. Translated by Lisa Richmond. Eugene, OR: Cascade, 2016.
———. *The Technological Society*. Translated by John Wilkinson. New York: Knopf, 1964.
———. *What I Believe*. Translated by Geoffrey Bromiley. Grand Rapids, MI: Eerdmans, 1989.

3

Oui et Non
Jacques Ellul's Dialectical Engagement with Karl Marx

BY Jacob Van Vleet

> Jacob Van Vleet is Assistant Professor of Philosophy at Diablo Valley College in California. He is the author of Dialectical Theology and Jacques Ellul *(Fortress, 2014)* and editor of Jacques Ellul: Essential Spiritual Writings *(Orbis, 2016)*. He is Vice-President of the International Jacques Ellul Society.

* * * *

> We owe to Marx the rediscovery of this central truth.... Marx brought history back to the light: not the history of historians, but history as we find it in the Bible: history filled with meaning, moving in a revealed direction, and culminating in an "apotheosis."... Here again Marx brings Christians back to revealed truth.[1]
>
> —JACQUES ELLUL

Needed: A New Karl Marx!

ELLUL'S UNDERSTANDING OF HISTORY and his interpretation of the Bible were significantly influenced by his reading of Karl Marx. Ellul first discovered the writings of Marx when he was eighteen years old. At the time, his father was unemployed and Ellul was supporting his family while a full-time student. He writes, "I was in the depth of despair.... I borrowed

1. Ellul, *Jesus and Marx*, 8–9.

Das Kapital from the library and started reading it. . . . I discovered a global interpretation of the world, the explanation for this drama of misery and decadence that we had experienced. The excellence of Marx's thinking, in the domain of economic theory, convinced me."[2]

It was from this moment that Ellul would go on to read Marx and engage with his thought for the remainder of his life. One of Ellul's earliest essays, first published in 1947, was entitled "Needed: A New Karl Marx!"[3] Often overlooked, this short work gives much insight into how Ellul read and applied Marx's thought throughout nearly all his writings, sociological and theological. Though Ellul's interpretation of Marx would evolve over the years, he was always sympathetic to Marx's thought, while at the same time harshly critical of any form of institutionalized or dogmatic Marxism.

A large part of Ellul's essay consists of a critique of two trends in Marxist schools of thought in postwar France. The first addressed those communists who believed that technological progress was the key to liberation. This group, sympathetic to Soviet Communism, argued that social problems will be solved once technologies are created to bridge economic inequality. For Ellul, new technology will never bring more freedom; it will only bring a certain level of material comfort to those with economic advantage. Ellul admits that new technologies can sometimes help the disenfranchised, but he is quick to caution that they will also create a new class of marginalized workers, thus alienating another sector of society. Ultimately, however, it is the illusion of technology's power to liberate which enslaves and leads people to place their hope in material fabrications rather than in each other.

Secondly, Ellul is highly critical of the utopian socialists. These Marxist thinkers believe that, through revolution or some other means, society as a whole must be restructured and reorganized in an egalitarian non-alienating manner. The final result will be small communes which utilize technology in a democratic manner. However, as we will see, while technology may provide the means for democratic participation, it also often results in the loss of individual freedom and democracy. According to Ellul, these utopians recognize the damaging consequences of the technological-industrial society, but they fail to see the underlying cause of the problem: technique.

In order to understand Ellul's conception of the political and his reading of Marx, we need to grasp the fundamental concept of technique. The clearest explanation of technique is found in *The Technological Society*, originally

---

2. Ellul, *In Season*, 10–11.

3. Ellul, "Problèmes de Civilisation II: On Demande un Nouveau Karl Marx," 360–74. This essay was translated into English by Marva J. Dawn and published in *Sources and Trajectories*, 31–45.

published in France in 1954 and appearing in English translation in 1964. In general, Ellul defines technique in two ways. First it is "the totality of methods rationally arrived at and having absolute efficiency (for a given stage of development) in every field of human activity."[4] In other words, technique is the ensemble of means which strives toward calculated efficiency in all sectors of society. We can see examples in educational, military, business, and even healthcare institutions, where we find an increasing worship of and advancement toward greater efficiency and productivity.

Second, technique is defined by Ellul as "the technological phenomenon" in the modern world.[5] The proliferation, inescapability, and reliance on technology and technological devices comprise this phenomenon. For Ellul, technique is simultaneously a mindset that strives for efficiency (the first definition) and the growth of technology (the second definition). They are interrelated and interdependent, each fostering the growth of the other. The ever-increasing domination of our lives by technology, whether with automobiles, airplanes, computers, televisions, or cell phones, grows and develops inseparably from our society's dependence on it. We can no longer live outside technology. It is an all-encompassing system. Thus, for Ellul, we live in a "technological milieu."

As a mindset, technique has in part been formed by the technological system itself. Our modern worldviews have been structured in a way that mirrors technology and its values. For example, the driving force behind each and every new technology is the quest for efficiency. Efficiency—and its corollary values such as competition and productivity—has become a determining value in a much broader way. Humans have blindly adopted this value, applying the standard of calculated efficiency to nearly every sector of society.

Ellul writes that neither the technological optimists nor the socialist utopians have properly diagnosed the malady of contemporary society: technique.[6] According to Ellul, Marx—not his followers—teaches us how to reach the correct diagnosis by providing us with a sound method—a method which his followers have misunderstood or neglected. "Marx was the only man," according to Ellul, "who grasped the totality of the social, political, and economic problems in their reality and posed correctly the

4. Ellul, *Technological Society*, xxv.
5. Ellul, *Technological Society*, 3–13.
6. For Ellul these two strands were continual trends within the larger Marxist discourse. Regardless of Ellul's accuracy in his criticism of these two groups, one can certainly see affinities between the technological optimists and thinkers like Mao Zedong, and between the Utopianists and thinkers such as Henri de Saint-Simon or Charles Fourier. See Ellul, "Needed!" 38.

questions [facing] the civilization of the nineteenth century."[7] Ellul goes on to point out that Marx was able to see clearly and "objectively" the "general conditions" which gave rise to inequality.[8] Marx's genius, for Ellul, lies in the fact that he saw the world both politically and spiritually. On the one hand, Marx recognized the interdependent relationship between economics and politics, and how each destroys democracy and leads to bondage. On the other hand, Marx, even as an atheist, was guided by a "spiritual" inclination that vehemently protested all forms of exploitation and dehumanization.[9] Marx was not content to sit back and analyze the political; he threw himself into protest and revolt in order to try and create a better society. "Marx is a good guide, for he too, let himself be led by his spiritual judgment of things."[10]

Ellul goes on to maintain that Christians should primarily glean two things from Marx. First, Christians need to see the world *realistically*. That is, they should not put hope in institutional religion, or in political or economic systems which have all proven to be inherently flawed. Second, Christians should *engage actively* with their local communities in order to make concrete changes. One of Ellul's favorite statements was "think globally, act locally." This is exactly what Marx did, and what Ellul argues Christians should do.

While Marx saw the world clearly and realistically, Ellul points out that Marx's analysis is not entirely sufficient today. In our present age, it is not capitalism which ultimately provides the conditions of exploitation, but rather technique: the primary driving force behind all modern economic and political systems. In capitalist, communist, and socialist states, technique provides the "structures" or "fundamental givens" which are the basis of today's society.[11] Recalling Ellul's definition, technique is simultaneously the growth of the technological milieu and the mindset which mirrors this milieu, worshipping efficiency and control.

Furthermore, Ellul argues that Marx also slipped into the error of "double historical extrapolation" or historical determinism.[12] Marx

---

7. Ellul, "Needed!" 38.

8. Ellul, "Needed!" 38.

9. When Ellul refers to Marx as "spiritual" he is not referring to religious belief. Rather, Ellul is pointing out that Marx was guided by ethically normative convictions and that he embraced a non-reductionist view of human nature. This is particularly evident in Marx's early writings, although Ellul argues that this humanistic inclination can be found in Marx's later "scientific" writings as well.

10. Ellul, "Needed!" 41–42.

11. Ellul, "Needed!" 42.

12. Ellul, "Needed!" 38.

believed that past economic conditions necessarily give rise to our present circumstances and that our present economic conditions will necessarily give rise to predictable future circumstances. Ellul believes that this sort of deterministic outlook is purely ideological, arguing that while there is a causal link between economics and historical outcomes, this is not guaranteed. In addition to economic forces driving history, Ellul believes that there is also a spiritual component to human history, the Word of God. The spoken Word is absolutely free and unpredictable—not determined in any way. Furthermore, the Word can radically change individuals, and by doing so change the course of history. For Marx, the future is predictable. For Ellul, this is not the case.

In addition to his error of historical determinism, Ellul asserts that Marx places too much emphasis on human labor as a source of human value and worth. "For Marx, work is the key to . . . human specificity, the means of becoming human, the source of value, the possibility of creating added value, the condition of history."[13] For Ellul, this "unexamined presupposition" is false. Work cannot give one's life ultimate meaning or value, even if it is unalienated or emancipated labor. Ellul writes, "In reality, this idea [the over-emphasis on work] appears in the seventeenth century as a very specific bourgeois ideology. It is a typical ruling-class ideology, since work considered as the supreme human value enables the ruling class to justify its domination as the class requiring others to work."[14] Marx accepted this ideology as many do today. In contrast, Ellul maintains that only through faith in the Wholly Other can one find authentic meaning, value, and hope—not in one's occupation or profession.[15]

Marxist Themes in Ellul's Writings: A Challenge to Christians

Despite these criticisms, Ellul regularly uses Marx's concepts throughout his work. In Marx's early works, especially his *Economic and Philosophic Manuscripts of 1844*, the term alienation is employed.[16] As a consequence of capitalism and increased industrial growth, alienation is a state of being

13. Ellul, *Jesus and Marx*, 81.

14. Ellul, *Jesus and Marx*, 81.

15. Ellul writes, for many today "work is thus identified with all morals and takes the place of all other values." Furthermore, if one is "successful" in business or the workplace, then one is deserving of esteem. Ellul, *Pour qui, pour quoi travaillons-nous*? 74.

16. For some key passages in Ellul's work on alienation see, *Ethics of Freedom*, 23–47; *Propaganda*, 169–82; *Money and Power*, 20–21, 78.

in which individuals are separated from their natural state of existence. For Marx, those who live in a capitalistic state are alienated from their loved ones, spending long hours in the workplace; alienated from themselves, losing their sense of creativity, purpose, and meaning; alienated from nature, spending the majority of their lives on an assembly line or in a factory or mill; and alienated from the fruit of their labor, never free to access the products or profit they generate at work.

Ellul largely agrees with Marx's theory of alienation, but sees alienation through a different lens. For Ellul, technique, not capitalism, is the primary cause of alienation in the contemporary world. Once one is continually enveloped in the technological milieu, one begins to mirror the values of technique: efficiency, power, and violence among others. When one's life is guided by these values, one will certainly become separated from authentic relationships with others, oneself, and nature.

Along with technique, propaganda is another primary cause of alienation today, says Ellul. Advertisers, politicians, and others employ psychological manipulation in order to control the masses. One of the largest components of this propaganda involves appeals to fear. Fear of terrorism, fear of financial insecurity, and even fear of immigrants, for example, cause a large segment of the population to live in constant anxiety. Living in a continual state of alarm and worry, year after year, leads to profound alienation.[17]

While utilizing Marx's theory of alienation as a component of his social criticism, Ellul also utilizes five broad aspects of Marx's thought to challenge his Christian readers.[18] The first of these aspects is the opposition to social injustice. For Marx, the goal of capitalism is to efficiently gain profit. This inevitably leads to the worship of money. When capital is valued over the health and well-being of individuals, injustice is necessarily caused. Humans and nature are exploited for monetary gain. Ellul believes that Christians should, like Marx, take these injustices seriously and work toward creating communities which abolish the worship of money and its deleterious effects.

Second, Marx always sided with the poor. Throughout his work, Marx fiercely advocates for the oppressed, the exploited, and the marginalized. Ellul agrees with Marx as one who stands with the poor. He states: "The poor person is indeed a person who requires a response; his very

---

17. Ellul, *Propaganda*, 169–82.

18. Frédéric Rognon insightfully explains these characteristics. See Rognon, *Jacques Ellul*, 211–33.

existence questions our lives."¹⁹ Furthermore, "It is the way the power of money works. Its most tangible form is what we call slavery, but we must realize that the poor person's situation is not much different from that of a slave. According to the Bible, it is extremely easy to slip from poverty into slavery. The purchase of a slave is the purchase not only of a body but of the whole person. Poverty also leads to the total alienation of the poor, an alienation which puts the labor force at the disposal of the wealthy, permitting the wealthy to impose their own law and conception of life, their own thought and religion."²⁰

The worship of money in our modern society leads to increasing selfishness and greed.²¹ Poverty and slavery are also common outcomes. Ellul argues that many Christians give lip service to the poor, but in reality they do little to help. Christians must work against those conditions and act as true advocates of the poor.

In addition, Ellul believes that Christians often lack coherence when it comes to praxis—uniting theory with concrete practice. Here he turns to a third aspect of Marx's thought: his articulation of the necessity of praxis. "For Marx, a sound ideology is possible only if it is related to a sound praxis."²² Throughout his life Ellul strove to live out his Christian faith in a practical, tangible way. For example, beginning in 1958, Ellul and his colleague Yves Charrier started an organization for marginalized young people, many of whom were living in poverty or on the streets. This organization gave free instruction in canoeing and kayaking, metal and wood shop, and even opened a beauty parlor for young women from the street to learn to become beauticians.²³ Ellul was actively involved in this work for many years, and it is only one example of how Christians might combine thought and action.²⁴

---

19. Ellul, *Money and Power*, 142.

20. Ellul, *Money and Power*, 78.

21. Ellul writes: "Without money, people are only paupers. They stand outside society and have hardly any place or function. We can almost say with Marx that in our world people exist only because of what they have, and when they have nothing, they do not exist. This is true in all societies, not only in capitalism, but the superiority given to material things in capitalism or socialism (it all comes to the same thing) makes this phenomenon more obvious and more severe in our time." Ellul, *Money and Power*, 141.

22. Ellul, *Jesus and Marx*, 115. Marx writes: "My *general* consciousness is only the *theoretical* shape of that which the *living* shape is the *real* community, the social fabric, although at the present day general consciousness is an abstraction from real life and as such confronts it with hostility. The activity of my *general* consciousness, as an activity, is therefore also my *theoretical* existence as a social being." Marx, *Economic and Philosophic Manuscripts*, 137.

23. Ellul, *In Season*, 119–20.

24. See Ellul, "Un club de prévention." In discussing his time working with

The fourth concept Ellul advocates is Marx's concern with "the concrete" rather than abstract theories, which is evident from his oft quoted statement: "The philosophers have only *interpreted* the world, in various ways; the point is to change it."[25] Ellul agrees wholeheartedly with this sentiment. In fact, Ellul maintains that many Christians have lost touch with the emphasis on the concrete found in the Bible. Ellul points to the prophets of the Hebrew Scriptures, who were not primarily concerned with abstract belief, but with actually aiding the widow and the orphan. Ellul writes: "The prophet is in effect the man who brings the Word of God to bear on the actual, concrete situation of man, his political situation."[26] And, "Man is thus summoned to participate in one or another of the actions of the Holy Spirit, in the totality of the work of God. He is summoned to provide the basis for the divine efficacy."[27]

Furthermore, Ellul argues that there is strong emphasis on the concrete in the New Testament, and points to two striking examples. First, the life and miracles of Jesus were largely concerned with a material nature—curing disease, restoring life, multiplying bread, and more. Second, Ellul maintains that the incarnation is a concrete embodiment of God's love for humanity and the earth—not an abstract one. He explains, "As for Jesus, a simple reminder of His incarnation betrays the dreadful error involved in a disembodied Christianity. One can also see in the Gospels the importance of daily life, the body, and the undivided unity of being. Thus Christianity has utterly betrayed the very essence of revelation by transforming it into religious spirituality."[28]

Along with other material emphases in the Old and New Testaments, the incarnation illustrates concrete involvement in our communities should be taken seriously. Ellul calls upon Christians to take concrete action, but unlike Marx, argues that they should resist employing violent means or putting their faith in merely human successes. In other words, Christians must refuse to engage in propagandistic techniques (psychological violence) or physical violence, and they should never try to calculate the success of their

---

marginalized youth in Bordeaux, Ellul emphasizes that he never tried to "moralize" or convert young men and women to Christianity. By providing free living arrangements and instruction on a variety of practical subjects, youth were able to stay out of trouble with the law and become less self-destructive. Ellul writes, "In reality, it is not the youth who are maladjusted to society, but our society which is maladjusted to human beings." "Un club de prévention," 77.

25. Marx, "Theses on Feuerbach," 123.
26. Ellul, *Politics of God*, 50.
27. Ellul, *Politics of God*, 135.
28. Ellul, *Jesus and Marx*, 8.

actions. Instead, Christians must live day by day, guided by the Holy Spirit, placing the final outcome of their actions in God's hands.[29]

Marx's ultimate aim was the well-being, cooperation, and thriving of community, and Ellul points to this as the fifth and final aspect of Marx's thought to challenge Christians. Ellul believes that today's Christianity is highly individualistic and has lost sight of the communitarian emphasis in the Bible. He writes: "Christianity has perverted Christian action by reducing it to a matter of individual conversion."[30] Instead of narcissistic individualism, Ellul urges us to return to the scriptures where we read "All who believed were together and had all things in common; they would sell their possessions and goods and distribute the proceeds to all, as any had need. Day by day, as they spent much time together in the temple, they broke bread at home and ate their food with glad and generous hearts," and "The whole group of those who believed were of one heart and soul, and no one claimed private ownership of any possessions, but everything they owned was held in common."[31] These are but a few biblical examples of the radical emphasis on community that Ellul begs contemporary Christians to recall and adopt.[32]

For Ellul, Marx was a prophet. Marx fought against injustice, emphasized praxis, and reminded his followers to reject narcissistic individualism. Marx also—and perhaps most importantly for Ellul—urged us to focus on the direction of history and to understand that we have a role to play in this unfolding drama. Ellul believes many Christians have lost sight of the meaning and purpose of history, and of the revolutionary responsibility they are called to take in this process. He emphasizes that Marx is crucial in regaining focus:

> When reading Marx, I was brought back to the social and political proclamations of the prophets.... It was Marx who convinced me that people in the various historical situations they find themselves, have a revolutionary function in regard to their society. But one must understand exactly which revolution it is; and in each historical period one must change, one must rediscover.[33]

29. Ellul, *Politics of God*, 136.
30. Ellul, *Jesus and Marx*, 8.
31. Acts 2:44–46; 4:32 (NRSV).
32. For an insightful commentary on these passages in relation to community in the primitive church, see Miranda, *Communism in the Bible*.
33. Ellul, *Perspectives on Our Age*, 13, 11. Ellul's deep knowledge of Marx is best seen in two books that have yet to be translated into English. The first, *La pensée marxiste*, is a presentation of Marx's thought and includes a detailed discussion of dialectic, materialism, labor and surplus value, capital, ideology, the state, and alienation. The

By first becoming aware of history and our unique place within it, we must then recognize our revolutionary function. For Ellul this may entail nonviolent resistance against a particular state or political entity. It will always, however, involve a fight against the myths and idols, the principalities and powers which continue to dehumanize individuals and destroy the earth.

Bibliography

Ellul, Jacques. *Ethics of Freedom*. Translated by Geoffrey Bromiley. Grand Rapids: Eerdmans, 1976.

———. *In Season, Out of Season: An Introduction to the Thought of Jacques Ellul*. Translated by Lani K. Niles. San Francisco: Harper & Row, 1982.

———. *Jesus and Marx: From Gospel to Ideology*. Translated by Joyce Main Hanks. Grand Rapids: Eerdmans, 1988.

———. *La pensée marxiste*. Edited by Michel Hourcade, Jean-Pierre Jézéquel, and Gérard Paul. Paris: La Table Ronde, 2012.

———. *Les successeurs de Marx: Cours professé à l'Institut d'études politiques de Bordeaux*. Edited by Michel Hourcade, Jean-Pierre Jézéquel, and Gérard Paul. Paris: La Table Ronde, 2007.

———. *Money and Power*. Translated by LaVonne Neff. Downers Grove, IL: InterVarsity, 1984.

———. "Needed! A New Karl Marx." In *Sources and Trajectories: Eight Early Articles by Jacques Ellul that Set the Stage*. Translated by Marva J. Dawn. Grand Rapids: Eerdmans, 1997.

———. *Perspectives on Our Age: Jacques Ellul Speaks on His Life and Work*. Edited by Willem H. Vanderburg. Toronto: House of Anansi, 2004.

———. *The Politics of God and the Politics of Man*. Translated by Geoffrey W. Bromiley. Grand Rapids: Eerdmans, 1972.

———. *Pour qui, pour quoi travaillons-nous?* Edited by Michel Hourcade, Jean-Pierre Jézéquel, and Gérard Paul. Paris: La Table Ronde, 2013.

———. *Propaganda*. Translated by Konrad Kellen and Jean Lerner. New York: Knopf, 1965.

———. *The Technological Society*. Translated by John Wilkerson. New York: Vintage, 1964.

———. "Un club de prévention." *Foi et Vie*, July 1989.

Marx, Karl. *Economic and Philosophic Manuscripts of 1844*. Edited by Dirk J. Struik. New York: International Publishers, 1964.

———. "Theses on Feuerbach." In *The German Ideology*, edited by C. J. Arthur. New York: New World, 1970.

Miranda, José. *Communism in the Bible*. Eugene, OR: Wipf & Stock, 2004.

Rognon, Frédéric. *Jacques Ellul: Une pensée en dialogue*. Geneva: Labor et Fides, 2007.

---

second book, *Les successeurs de Marx*, is primarily a presentation of major Marxist thinkers (Jean Jaurès, Georges Sorel, Eduard Bernstein, Karl Kautsky, Vladimir Lenin, Rosa Luxemburg, Georgi Plekhanov, Radovan Richta, and Ota Šik) and an explanation of the divergent paths of Marxist thought. Both books were compiled from courses Ellul taught at the Institute of Political Studies at Bordeaux between 1947 and 1979.

4

The Political Theology of Jacques Ellul

BY David Gill

> *David Gill (BA, UC Berkeley; MA, San Francisco State) earned his PhD at the University of Southern California with a dissertation on* The Word of God in the Ethics of Jacques Ellul, *subsequently published as the first of his seven books on theological or business ethics. He spent a sabbatical year, 1984–85, and several summers in Bordeaux, meeting with Ellul and many Ellul scholars, family, and friends. In June 2016 he transitioned from a forty-year career as a business school and seminary professor of ethics to life as a writer. Since 2000 he has served as founding president of the International Jacques Ellul Society and a founding board member of the Association Internationale Jacques Ellul.*

* * * *

JACQUES ELLUL'S HISTORICAL AND sociological writings on the state and politics are, by themselves, extraordinarily deep and illuminating. Add to that his theological perspectives and the richness of his political analysis becomes all the greater.[1] In his theological writings Ellul is usually speaking *from within* the community of Christian faith and *to* that community of faith. Nevertheless, whether readers personally share Ellul's faith or not, they regularly testify to the value of engaging this theological side of Ellul. It reminds me of Matthew's account (chapters 5–7) of the Sermon on the Mount. Jesus saw the crowds, then went up on the mountainside, his disciples followed, sat down, and he taught them the distinctive character, values, and ways of the kingdom. But—surprise!—"when

---

1. The best introductory essay on Ellul's sociology and theology of politics is Troude-Chastenet, "Jacques Ellul et la politique," 33–59. Troude-Chastenet's interviews with Ellul are also an essential and fascinating source. See *Jacques Ellul on Politics*.

he finished these words, the crowds were amazed at his teaching" (Matt 7:28).[2] So with Ellul: what he addresses to his Christian community finds resonance and often even "amazes" a much wider audience.

I will sketch out Ellul's political theology in three basic steps. First, what does biblical theology have to say about *political/governmental institutions*, at the center of which is the *state*? Second, what does it say about *political activity* in relation to the state, party, or other political institution? Third, beyond making us "aware" of the truth and reality of the state and politics, what does Ellul advise the church actually to do in relation to this domain?[3] In typical Ellul fashion, there will be no system, no comprehensive theory, no dogma, no abstract principles—but rather insights, perspectives, stances, and suggestions. But if followed, these could truly rock both church and world. I will do my best to explain and illustrate Ellul's ideas as we go and then close with some summary observations.

The State in Light of Biblical Theology

In a hundred pages of his 1954 classic *La Technique, ou l'enjeu du siècle* ("Technique, the Stake/Issue of the Century," translated and published in 1964 as *The Technological Society*), Ellul described how the modern state, east and west, liberal or fascist, no matter the rhetoric or ideology of its leadership, has been taken over by *technique*, by bureaucracy and the ruling value of efficiency, the "one best way" as defined and measured by technicians.[4] In his 1965 major work *L'Illusion politique* (translated and published in 1967 as *The Political Illusion*), Ellul describes the illusory quest for "control of the state" when this bureaucracy and administrative technique are dominant.[5]

Politics is conducted in "the ephemeral," the world of images, and has little impact on actual structures, policies, and decisions. Ellul paints a bleak picture of out-of-control government, constantly expanding to absorb all

---

2. Ellul was a hard-core Barthian who had no time for any kind of "natural theology" apart from biblical revelation. I am a little less "hard-core" in that I think, like St. Paul (Rom 2:14–15), that everybody, not just Jewish and Christian believers, has in some mysterious way the "law written on their heart" and a "conscience" that bears witness to what is true. I think this explains why readers "resonate" with the insights of Ellul's theology even without sharing his personal faith.

3. Throughout this essay, by "church" I do not mean the institution but simply Christian individuals or groups.

4. Ellul, *Technological Society*, 228–318.

5. Ellul, *Political Illusion*, 136–62.

aspects of the life of citizens, constantly subordinating and reducing human life to numbers and statistics, to a calculation of means.

> The state in any country, no matter what may be its form, whether democratic or dictatorial, new or popular, is in fact the chief danger known to humanity, whether from the material standpoint or the spiritual standpoint.... Every modern state is totalitarian. It recognizes no limit either factual or legal. That is why I maintain that no state in the modern world is legitimate. No present-day authority can claim to be instituted by God, for all authority is set in the framework of a totalitarian state. This is why I decide for anarchy.[6]

Of course, Ellul acknowledges, there are legitimate and unavoidable governmental functions that must be carried out (maintaining order and tranquility, building roads and infrastructure, preventing business fraud, and running a postal service could be on this list) and those who work on these matters, Christian or otherwise, should do them well. The problem is that the modern state recognizes no limits and is ever-growing and ever-expanding, seizing power over more and more of our lives.

From a sociological perspective, the modern nation-state is increasingly serving as the "sacred," the center of life. "The state is the ultimate value which gives everything its meaning. It is a providence of which everything is expected, a supreme power which pronounces truth and justice and has the power of life and death over its citizens."[7] The nation-state becomes an idol, the center of value, the source of inspiration, awe, and even adoration, the demander of sacrifice, the dispenser of salvation, enlightenment, healing, and meaning. "What makes it sacred is not that it sets itself up as God but the fact that the people accept it, live it, and look upon it as the great ordainer, the supreme and inevitable providence."[8]

While this is true on a mundane, day-to-day level, we should also think of the Nazi rallies and their truly "religious" grip on the people. Not so different are today's Trump rallies with their emotional chants of "America First" and a sycophantic priesthood bowing and scraping before him, shouting "Amen" as he takes credit for all things positive and finds scapegoats for all things negative. "Praise you, oh Mr. President, what an honor to be in your presence, probably the greatest president of all time!" Please note, of course, that while Trumpism is an especially garish example, this shows up in Pyongyang, Manila, Moscow, and many other places around the globe.

6. Ellul, *Ethics of Freedom*, 396.
7. Ellul, *New Demons*, 80.
8. Ellul, *New Demons*, 81.

Alongside this rather bleak sociological analysis of the modern state, Ellul's theological studies are darker still. The topic comes up in many of his writings but his 1973 article "Rappels et réflexions sur une théologie de l'état" and his 1975 study of the book of Revelation might be the two best starting points. To begin with, the church must understand that the state is legitimate and necessary, that it "finds it foundation in the will of God, that it has meaning in relation to the work of God, that it has received its authority from God, that it has a vocation."[9] While that is true, Ellul cautions against emphasizing this theological point too loudly because the state might take it as justification to seek even more power, now in the name of God!

Ellul argues that the whole Bible displays a lack of interest and guidance on the particular form a state may have. Christians have sometimes advocated monarchy as a sort of human imaging of God's kingly status and rule. Ellul points to God's warnings to Israel about their desire for a king "like the other nations" (1 Sam 8). If they depart from the essentially a-political, limited-institutional rule of God through judges, priests, and prophets, and demand a monarchy, they will soon be subjected to military service, taxation, and all sorts of domination and exploitation by their king. Not a good idea as it turned out.

More recently some Chistians have defended a communist state by stressing biblical passages about the importance of the poor and evils of wealth in God's eyes. Didn't the early church of Acts 2 and 4 move away from private ownership and share "all things in common"? Still others advocate for democracy by appealing to the universality of the image of God and the gifts of God in all men and women. The fact is there were many different political institution models in the time of Jesus and throughout the whole biblical epoch. Yet we are given no guidance or opinions in Scripture on preferred constitutions. This doesn't mean that all forms are equal. Ellul prefers democratic institutions, he says. But we cannot theologically justify making one choice or another into some kind of divine absolute.

One thing on which all of Scripture insists, Ellul writes, is that the political power must be *limited*. The famous episode of Jesus and the question of paying taxes illustrates the limits of state power (Matt 22:17–22). The money bears Caesar's image and it is within his right to claim a tax. But human beings bear God's image and must not be claimed by Caesar. The wicked kings of the Old Testament are guilty of pride, of exalting themselves, of extending their power, of using religion for political gain (the "sin of Jeroboam"), of even thinking they are gods. Ellul argues that the state must be limited in its

---

9. Ellul, "Rappels et réflexions," 154. My translation.

scope and reach, but also in its duration. No state is eternal. All are temporary and relative to a given historical situation.

Ellul also sees a profound political lesson in the three temptations of Christ (Matt 4:1–11). In one temptation, the Devil took Jesus to a high mountain and showed him "all the kingdoms of the world and their glory." Satan offered these kingdoms to Jesus if he would just bow down and worship him. Jesus refused, of course, but Ellul remarks that Jesus did not dispute the fact that Satan had control of the nations!

We should not forget the role of the state and political power in crucifying Jesus. This was not just because of religious opposition. Pontius Pilate was the governor, not the high priest. Later in the New Testament St. Paul uses the vocabulary of *exousia*—the "principalities and powers"—to refer to the state.[10] These "powers" represent a kind of mysterious, collective, rebellious, spiritual reality that corrupts institutions, creates idolatry, and fights against people. St. Paul advised the young Ephesian church that life's struggle was not so much against flesh and blood as against such powerful spiritual forces (Eph 6:12). The political *exousia* are ordered and used by God in some way (Rom 13:1–7). They have a legitimacy. But even so, there is no invitation or suggestion by Paul that Christians should join the powers in their activity (that is, as a part of their Christian discipleship).

In the final book of the Bible, the Revelation/Apocalypse, two vivid apocalyptic figures represent for Ellul the essence and core truth about the state in human history: the red horse (among the four horsemen of Rev 6:1–8) and the first (of two) "beast(s)" (from Rev 13).[11] Ellul argues that the primary forces (and meaning) of human history are disclosed in the symbolism of the "four horsemen of the Apocalypse" galloping across history. The first historical force is the white horse representing the Word of God, which is Truth incarnate in life. The second force, the red horse, represents the state with its power of wielding the sword and waging war. "This horseman is at the same time an *exousia*, the state itself, and the political scourge of war."[12]

Later in the book of Revelation, the state political power is described as the first of two ominous beasts. Ellul comments:

---

10 Ellul, *Ethics of Freedom*, 152–56.

11 Ellul, *Apocalypse*. The four horsemen are discussed pp. 147–56; the beasts are discussed pp. 92–99.

12 Ellul, *Apocalypse*, 149. The third, black horse represents the economic power (for good or ill); the fourth, pale horse represents the power of death/pestilence throughout history.

It is denoted a political power which will be exercised over all people. Not an organization . . . but the Power itself in its political face (today, the state). It holds the sword and brings death. It decides captivity and imprisonment. But in addition, in regard to the Church and Christians, it holds power and conquers them upon both the material and spiritual planes; for this state is not only force and power but also seduction and the capacity to make itself adored. . . . Now this power of the state is given to it by the Dragon . . . which is to say that the power of the state is not of the natural, naturalistic, sociological order; it comes from the power of chaos, from the destroyer; as admirably organized, regulated as it is, it always expresses chaos. The more the state order reigns, the more the disorder of the Dragon prevails. Such is the message of the Apocalypse on the state. And precisely because it expresses a spiritual power, people, who feel it deeply, worship it.[13]

The Dragon is the divider and liar in Rev 12. Not very promising!

So theologically, as well as sociologically, the modern state for Ellul is not something fundamentally positive. It is a dark, menacing, deceptive, violent, monstrous force marauding across history and the planet, pretending to be God but acting like the Devil. It is no beneficent (or even ambivalent) "order of creation." By contrast, even in Ellul's theology of the city, big bad Babylon may dominate and overshadow human history but there is still the flickering light of the cities of refuge, historical Jerusalem, and the eschatological promise of the New Jerusalem after the final judgment of Babylon the Great. There doesn't seem to be a parallel eschatological "New Political Dog" to emerge after the final judgment of the Political Beast.

Augustine, Aquinas, Calvin, Luther and others of a more positive bent like to see political organizations and institutions as at least potentially hopeful, positive, creative structures for a civilized life together from Eden onward. They may be failing, imperfect, and even power-hungry but that doesn't mean there could be absolutely no hunger *also* for peace or justice. Among other possible biblical resources is there really nothing to learn about political organization and law from the way Israel was set up? Or from the stories of the Judges? Or the outlines of the eschatological Kingdom of God? Nothing can be simply or wholly transposed from the Bible to our place and time, of course. But must the red horseman and the beast alone provide our core lessons? Just asking!

---

13. Ellul, *Apocalypse*, 94–95. The second beast, Ellul writes, represents propaganda, inducing the people to worship the first beast, state political power.

Political Activity in Light of Biblical Theology

Political activity, Ellul says, is *important* but it is not the main activity of human beings. Politics is of *relative* not absolute importance. And just what is political action about? "The real problem is that of active participation in real political action, that is, the *discharge of a directive function in a party or a state organism. In this alone is one engaged in politics*. The rest is a matter of opinion, obedience, or debate, but it is in no sense politics."[14] "The term 'political' must be taken here in its precise and restricted sense, i.e., with relation to the state and not just to any power, or just any social activity. Max Weber's definition is both classic and excellent: 'Politics is the leadership by a political body called the state, or any influence exerted in that direction.'"[15] Ellul's perspective represents a narrowing of the meaning of politics that restricts it to the official or recognized apparatus of party and state. Can an action be political if it remains outside the official legal and civil institutions but affects the lives and behaviors of a community of citizens? Ellul implies "No" in his definitions.

The political illusion, Ellul describes at length, is (1) that we can control the state (we can't; it is controlled by the technicians and bureaucrats or, more accurately, by Technique understood as the whole complex of tools, processes, and values aiming at the "one best way"); (2) that we are able to participate in real politics, i.e., the control and direction of state and party institutions (no, we are drawn into the ephemeral, superficial sound and fury of political debates and activities but this is not consequential at a deeper level); and (3) that there are political solutions we can achieve (no, the technical apparatus of the state operates with but one solution in mind: greater efficiency; our values are excluded and, frankly, the most important problems of human life are not amenable to political solutions in any case).

Ellul's theological perspective on political thought and activity is best approached first in his 1966 study of II Kings, *The Politics of God and the Politics of Man* and his 1978 essay "Modern Politics, Realm of the Demonic" (which also was included in his 1980 book, *Living Faith*). Ellul writes that II Kings is all about what is "political in the narrow sense; the problems in most of the texts are political. . . . The Second Book of Kings is probably the most political of all the books of the Bible. . . . Above all we see here politics in action and not just in principle. . . . Politics is not just simply a human

---

14. Ellul, *Politics of God*, 14, emphasis added.
15. Ellul, *Political Illusion*, 15n6.

action of no concern to us. It may be that politics is the kingdom of the devil, but this certainly concerns us as Christians.[16]

As he moves into the stories of II Kings, Ellul insists "God's action in politics will continually have for us the appearance of vocation, appeal, and address, and then judgment, outburst and wrath. It will continually have for us the appearance of grace, of timid approach, of liberation, then of rigor, of inflexibility in attaining its specific end, and sometimes, if rarely, of a miracle which intervenes to overthrow the course of events, of history, and of life.[17]

Ellul first turns to the story of Naaman, a warrior leader in an enemy neighboring state, who has been afflicted with leprosy. Naaman turns to another political leader, the king, to seek help but God chooses to heal him through a humble servant girl and a prophet. Despite the healing, relations between Israel and Syria do not improve and soon war breaks out. Ellul sees this episode as an indication of God's simultaneous interest and disinterest in the political world and of his characteristic choice to work on human problems through humble means. God will save Samaria not through bombs, elections, the courage of soldiers, the skill of generals, or the politics of a king, but through rejects and outcasts, through the Word, through a miracle.

So too, our political passions, military and revolutionary storms and dramas are more noise than substance. God laughs at our overly serious and pretentious strutting around. Our efficacy may appear a failure in the world's eyes but God measures our actions by faithfulness rather than quantifiable success. "The efficacy we seek . . . is the efficacy of event as opposed to institution, of tension against the accepted line, of nonconformity. In sum, it is an efficacy which stands opposed to that of the world. Yet it is no less real. It is the efficacy of heretics and sectarians."[18]

Many of these kings of Israel and Judah are charged by the prophet with the "sin of Jeroboam"—using and exploiting God to enhance the state. Throughout the stories of II Kings, the role of the prophet is huge, articulating a word, often a warning, from God to the political leadership and to the people.

> Elisha is the visible and active presence of God himself. At every instant he carries consolation for the poor and afflicted. . . . He is the sign and proof and witness that God has not abandoned his people. . . . Even if the prophet does not change events, the fullness of God is a sufficient answer. The same seems to be true

---

16. Ellul, *Politics of God*, 13.
17. Ellul, *Politics of God*, 22.
18. Ellul, *Politics of God*, 141.

of the church, the body of Christ. Perhaps it does not have to intervene directly in politics. But its presence is enough to make the worst disasters endurable.[19]

Ellul sees the story of Rabshakeh as an account of the fundamental politics of the world, exhibited in realism, calculation, force, threats, and propaganda. This sets the stage for good king Hezekiah, one of the leaders attacked by Rabshakeh. Hezekiah responds to the political crisis by tearing his clothes, retreating, and repenting in the temple before God. He has no political or military answer for Rabshakeh. He can only retreat, repent, and call on God. In the story, God acts in a miraculous, powerful way.

As long as politics is relative, Ellul says, Christians can act. But when it becomes the beast, and succeeds in displacing God, we may be left with little alternative but to repent and pray like Hezekiah.

> So long as the political debate is within the relative sphere, the Christian can play a part with his own proper methods and forces and his own responsibility. . . . But when one of the powers claims to be God, to embrace the totality of human life, to give total meaning to action, history, and life, no relation is possible, not even that of conflict, for one absolute claim can only be met by another absolute claim. In the presence of one who claims to be God, victory can go only to another who claims to be God.[20]

Ellul's final chapter in *Politics of God* is a "Meditation on Inutility" (uselessness). What are we to make of the stories of II Kings?[21] Certainly one lesson is how rare it is to find a good king, one who observes the law, who does not try to use God for his own political purposes, and who upholds justice for all the people. So don't put your trust in political leaders! Watch and pray for God to act through humble and unexpected means. What appears to be ineffective and useless now may be effective in the long run and on the deepest level.

Thus far, Ellul is pointing mainly to the pretentiousness and arrogance of political leaders and to the apparent uselessness and illusion of political activity. In his essay "modern politics, realm of the demonic" he ratchets up the critical language.[22] This deserves quoting at length:

---

19. Ellul, *Politics of God*, 90–91.
20. Ellul, *Politics of God*, 170.
21. Walzer's *In God's Shadow*, underscores much of Ellul's perspective on politics in the Old Testament.
22. Chapter 21 of Ellul, *Living Faith*, is almost identical in title and content to his 1978 article in Olivetti, *Religion et politique*. My quotations are from the *Living Faith* version.

> If evil has piled up upon evil, if the tide of danger is rising, the reason lies in politics and nowhere else. Politics is the contemporary image of absolute evil. It is satanic, diabolical, the home base of the demonic. And when I say politics, I am not pointing at the state—that's another problem again. . . .
>
> No, the issue is politics itself, whatever form it may take, whatever its objectives, doctrines, methods, social roots, intentions or rationales. I am talking about politics in the concrete, as put into practice by the political world. . . .
>
> Politics is the acquisition of power: the means necessary for getting it, and once you have it the means for defending yourself against the enemy and so holding on to it. But what does one use it for—for goodness and virtue? No, one uses it for power; it's an end in itself. And that's all there is to politics. All the fine talk about politics as a means of establishing justice, so forth and so forth, is nothing but a smokescreen that on the one hand conceals harsh, vulgar reality and on the other justifies the universal passion for politics, the universal conviction that everything is political, that politics is the most noble human activity, whereas it is really the most ignoble. It is, strictly speaking, the soure of all the evils that plague our time. And when I say that it is diabolical and satanic, I mean those adjectives literally.[23]

The thirst for power is at the heart of all politics in Ellul's view. As a Christian, he sees this as a total contradiction to the God of Scripture and Jesus Christ. As harsh and total as Ellul's condemnation of politics may seem, we do have to notice how today's candidates for office will usually do anything and say anything to get into office, then to stay in office. All talk of saving future generations of Americans from onerous national debt running into the trillions of dollars goes right out the window when casting a vote to the contrary might increase one's odds of re-election.

Then there is the role of propaganda and lying in politics. "In our time the father of lies speaks through propaganda, which engenders passion and false clarity, burning commitment and inner alienation." We can see the "radical mendacity of the father of lies, who is literally and totally incarnate today in politics."[24] The demonic character of politics is further illustrated by the names "Satan" and "Devil." "In the Bible Satan and the devil are not identical. . . . The devil is the one who sows discord through seduction. Satan is the accuser. . . . Wherever accusation occurs . . . Satan

---

23. Ellul, *Living Faith*, 234–35.
24. Ellul, *Living Faith*, 240–41.

is at work.... The satanic is the pure distilled essence of the political.[25] Again, Ellul's condemnation sounds so extreme. But can anyone look at contemporary American political leadership—with its extreme partisan divisions, its daily litany of flat-out lies, its accusation and blaming, its divisive impacts—and not admit "he's got a point"?[26]

Again, I would first of all affirm Ellul's basic theological insights here on political activity. But I would also see some potential counterpoints of a more hopeful type. Without going into detail, I refer to Melchizadek the peaceful king of Salem, to Moses as political leader, to Judge Deborah, to King David's positive example (not just his big mistakes), to Joseph and Daniel working in political administrations in Egypt and Babylon. There are pagan kings and political rulers who do justice and are recognized as such in the Bible. Politics is not all a seething cesspool of violence, power-grabbing, lies, and theft without a hopeful, redemptive counterpoint.

Christian Political Responsibility and Opportunity

Ellul's experience after challenging Christians to be a revolutionary presence in the modern world in his 1948 classic, *Presence in the modern world*, was frustration and disappointment. After the end of World War Two, the launching of the World Council of Churches, the mobilization of the church to help with the reconstruction of Europe, and the deconstruction of the European empires, the church produced a plethora of position papers and official statements on political issues, but these were, in Ellul's view, often naïve, poorly researched, and frankly divisive in the church as well as the world. How might we get our bearings again?

---

25. Ellul, *Living Faith*, 241. While I agree with Ellul's basic argument here I have to push back on the terminology. Contra Ellul, Rev 12:7–9 suggests that the Devil and Satan are one and the same. Satan is the Hebrew term, Devil is the Greek translation of Satanas. In Greek "Devil" (or think of the Spanish "diablo") comes from *dia* (across) and *ballo* (to throw) suggesting "to hurl across" an accusation in a court room or debate. *Satanas* is the older Hebrew term meaning essentially the same thing. These accusations based on lies, falsehoods, and misrepresentations deceive, alienate, and divide people. The Revelation text also suggests that this evil power manifests sometimes as a ferocious Dragon, other times as a quietly slithering snake. Comprehensive and powerful.

26. Interesting that Ellul writes, "I have done quite a few studies on technology, but I've always been careful not to say that it was demonic or diabolical." "The structure of contemporary politics corresponds point for point with the structure of the demonic." But technique/technology, economics, science ... these do not fit the description. Ellul, *Living Faith*, 247.

In his *Ethics of Freedom*, Ellul outlined a three-part Christian ethics he planned to write. The first part would be the ethics of *freedom*.[27] In response to our *hope*, God sets us on a quest to experience and demonstrate freedom even in a world of necessity. What might it mean to act in and for freedom in the political arena and in relation to today's nation-state? The second part would be the *ethics of holiness*. In response to our *faith*, God sets us on a quest to experience and demonstrate *holiness* (meaning both "purity" and "otherness/distinctiveness") in the political realm and other aspects of life. If the holiness angle separates us, the third part, the *ethics of relationship*, responding to our *love*, sends us back into communion and partnership with others.[28] We can almost guess what Ellul might say about being distinctive and "wholly other" in the political domain.

But what does it mean to live out our *freedom* in politics? Ellul says that there are no abstract principles, no system here. "I will begin . . . with the ethical situation, e.g., the Christian as a free man in politics."[29] Christians must not merge into either Left or Right political traditions—nor can they adopt an apolitical position that shuns all politics. The fact is that with the growth of the modern state (and the decline of the church and other institutions), all of us are *already* deep in politics. There is no real escape. The question is just how we will exercise our freedom in today's political context?

Ellul has been very reticent to advise us on what we should actually do in light of the truth and reality he describes. "I refuse to construct a *system* of thought, or to offer up some Christian or prefabricated sociopolitical solutions. I want only to provide Christians with the means of thinking out *for themselves* the meaning of their involvement in the modern world."[30] Why this reticence? For one thing, despite various continuities, the world is always changing and so the advice for today may not fit tomorrow. Local culture and circumstance also vary across the globe. What is appropriate for Berkeley won't meet the challenges villagers face in India. Also we, the readers, the political agents, have very different gifts, experiences, passions, vocational callings, and opportunities. My agenda and strategy can not be the same as

27. The section of the 1976 translation and publication of *The Ethics of Freedom* focused specifically on politics and "the sovereign" (369–98) was based on a manuscript Ellul later edited slightly and published in 1984 as *Les Combats de la Liberté* (100–133). My quotations are from *Ethics of Freedom*.

28. Jacques Ellul's ethics of holiness has now at last been transcribed (by son Yves Ellul) into a 1,000-page typescript and we remain hopeful of seeing it published sometime in the next decade.

29. Ellul, *Ethics of Freedom*, 369. Note the "existential" feel of Ellul's thinking. Like Kierkegaard, we stand in this moment before God and this context requiring a decision. Will we be faithful? What is faithfulness at this moment?

30. Ellul, "From Jacques Ellul," in Holloway, *Introducing Jacques Ellul*, 6.

yours. And finally, Ellul's Kierkegaardian view of God as alive, dynamic, and Wholly Other always inhibits any formation of a permanent, detailed set of guidance for politics, ethics, or other aspects of life.

Nevertheless, all is not chaos and uncertainty. With the preceding cautions and qualifications in mind, it is possible to sketch out some pointers toward a response to the reality of the state and politics. Ellul provides at least passing hints and counsels for a Christian response in many of his writings but the most extended and explicit discussions can be found in *False Presence of the Kingdom*, *The Ethics of Freedom*, and an essay "Thèses sur foi chrètienne et politique." I will distill and summarize seven key points from these three statements made in 1963, 1976, and 1980.

1. Keep church and state separate. The first item on the agenda is to resist collapsing the radical divide between church and state. It is a mistake, Ellul writes, to make the church political, i.e., as one more interest group among others playing the political game, using the same political tactics and means as others. This corrupts the specific nature and calling of the church. It trades away the capacity of the church to bring in a word and a perspective from outside, from the God who doesn't just love one party or country but who loves all humankind and all nations equally. Bringing national flags, voter guides, and political candidates into the pulpit is almost always a terrible thing for the church. We need powerful and prophetic preaching and teaching on biblical stories and on biblical values such as justice, kindness, generosity, the treatment of children and foreigners, for example. But we should stick to the text and let the parishioners figure out the concrete political implications. It is a terrible betrayal of the faith to use it for political purposes—and to alienate and drive away those of an alternative political sensibility.

As bad as "politicizing the church" is the reverse movement of attempting to "Christianize the state." The state is incapable of walking in the way of Jesus or (contra John Calvin) governing by the Decalogue, which is the *covenant* between a redeemed people and their Lord, not a secularizable *constitution*.[31] It is ignorant and irresponsible to ask the state to be "Christian" (or Muslim, Jewish, Hindu, etc., for that matter) when it lacks the spiritual, guiding, and animating core of the faith. With Max Weber, Ellul believes the state depends ultimately (though not exclusively) on force, on the sword. But Christians are called to live by an ethic of

---

31. The recently defeated candidate for US Senate, former judge Roy Moore, became famous for insisting on a monument to the Ten Commandments in his court house. My strong objection was that his energy would be much better and more appropriately spent getting the Ten Commandments into the *church!* Ignorance of the Decalogue and its rich guidance is a plague in the church.

non-power, and to a life of service, not domination. Misguided attempts to make the USA a "Christian nation" are doomed to fail, alienating those who think otherwise (including lots of Christians), and bastardizing the "civil religion" parading as Christianity. Both church and state are deeply corrupted and subverted by any attempts to blend them together or to subordinate one to the authority of the other.[32]

2. Desacralize, demythologize, relativize. Second, there is the "stern duty" of relativization, desacralization, and demythologization. Relativizing politics does not mean it is devoid of importance.[33] No, it is important to run cities well, to promote better laws, to optimize public services. But if we can reduce passions and noise and division in public debate and forge proposals and agreements that help things to get marginally better, we should do so. But let's not treat politics as having absolute, ultimate value and importance. It is of *some* importance.

If the nation-state has become sacred and an idol, then it must be desacralized and dethroned.[34] How do we do that? First of all, we withhold our praise, our adoration and awe, and also our "prayers" (requests for help). We profane the (state) idol by treating it as ordinary, laughing at it, and "taking its name in vain." Just as we should not profane or take in vain the name of a true God, we should do the opposite with a false one. We must reject any religious exaltation of the nation-state with its chants of "America First" or "Deutschland Über Alles." We must not get choked up and have goose bumps when national hymns are sung and we are pressured to piously salute while standing at attention. But neither do we judge and condemn the state for failing to deliver as a god (it is incapable of this). Rather we judge it for its pretension and false promises. It is not worthy to be our God and it is incapable of being our Lord or our Savior. This is the "no" with which we must begin.

3. Be present in the political world. We work to keep the state and the church separate and we work against any idolatry and sacralization of the state. But the third point is that Christians must be present not only in the church but the political arena. Ellul constantly and passionately argues that the Christian life is one, unified, experience. It is 24/7. No dualism of sacred and secular, church-life and social-life, is permitted.[35] Christians are called to be present in the world, including its politics. In fact, they are implicated

---

32. Ellul, *Subversion*, 113–36.

33. Ellul, "Thèses," 194.

34. Ellul, *New Demons*, 222–28. In this case, we are talking about state and politics as *substitute* gods and religions.

35. There is a dualism of *institutions* (separation of church and state) but not a dualism of sacred and secular in our life, values, and thinking.

in the world's politics whether they know it or like it or not. The first act of freedom in politics is "choice and presence."[36]

This choice of where to be present, Ellul says, is not a particularly Christian or theological choice. It has to do with where and among whom we live and work, or with an affinity for particular people or positions, with personal taste or circumstance. It is actually *desirable* to have thoughtful Christians present in virtually all groups and all parts of the political landscape. Why? In order to have representatives of God's word on love and justice in each and all of these constituencies, not just all in one or two "Christian" groups. It is also important so that the unity of Christians across party divisions can be an exemplary and hopeful witness that political difference need not radically, permanently divide us. Too bad this is so rare!

4. Be an ambassador representing a "third way."[37] Here is where a critical point is made: in every case Christians must be present as bearers of the distinctive Word and values of God.[38] They have a prophetic witness to bring, speaking for truth and honesty, for the poor, oppressed, ignored, and those in bondage. Wherever Christians may choose to be present, their role is *never* to be an "amen corner," just blessing, endorsing, and "pouring holy water" on whatever their political colleagues and peers are doing and saying. Christians are always ambassadors from the "kingdom of God," not cheerleaders for an existing political interest group. It does no good for Christians just to assist their party's propaganda efforts and ratchet up the volume and heat.

For this "third way" advocacy to happen, the starting point is always careful study of the revelation, the Word of God in Scripture. This foundation alone is what enables Christians to "have something to say" that just might "bring about the event instead of trailing it, submitting to it, explaining it."[39] Because of its foundation in Scripture (i.e., rather than in a constitution or philosophy or assessment of the facts), the church stands in a permanent tension with any political authority. While the message cannot be an unqualified "Yes" to the state or party, neither is it a simple rejection and negation. It is "other"—a message of love and justice. And it is not just

---

36. Ellul, *Ethics of Freedom*, 375–81.

37. "Third way" is my choice of terms (not Ellul's) to suggest thinking "outside the box," beyond the usual polarized options.

38. Ellul, *Presence*, 26–29. Christians are *ambassadors*, representatives of another country, another Sovereign, another perspective, orientation, and set of values. Their role is revolutionary, against the tide, whichever country, whichever state, whichever party, organization, movement, or role they happen to occupy.

39. Ellul, *False Presence*, 182.

about ideas and content—the message must be incarnated, not just talked but walked, if Christians want it to be heard.

This is more than just passing resolutions or carrying signs at a rally. Ellul challenges Christian ambassadors to "speak to the sovereign."[40] The Bible focuses on the personnel, the people leading the state, not on the state as an abstraction or institution per se. Dialogue with the king, not with the monarchy, with the senators, not the Senate, with the people. "In our own time when we remember that the people is now the true political sovereign. It is to the sovereign that we must speak."[41] Who is the person with political authority in this or that situation? It requires a deep understanding of those to whom we speak. Go talk to the principal or the school board member—don't just address the institution. "What is needed is that power should be confronted by someone who uses other means, who says something other. . . . It is essential for the political health of the nation that power should not merely be in encounter with itself."[42]

This "third way" message stands in creative tension with ordinary politics and propaganda. Such tension is not a negative thing but something altogether critical to the health of both church and state. Resistance builds strength in physical, psychological, intellectual, and relational terms—and in politics as well. Opposition and tension are essential to both church and state. In fact, when the party in charge changes, Christians must shift their focus to challenging the new authorities in power. Christians should be a "permanent opposition."

Part of the impact of this strategy is to "open up the world." The tendency is for the modern state to become totalitarian (whether with kid gloves or brass knuckles, *Brave New World* or *1984*), taking over the whole of life, including its religious and spiritual dimensions. The church must not accept any colonization and domestication by the state but rather cultivate its own distinctive lifestyle and thought patterns, its own authority and message based on its relationship with God and the Bible.

5. Focus especially on five concerns. Of course, Christians should lend help along the way to their political or governmental colleagues and do their job well if they are part of the political organization. "Christians have to set an example. They have to show that these technical (and not ideological) tasks . . . and that these relative ends must be pursued seriously."[43] But Ellul suggests concentrating on five other matters.

---

40. Ellul, "Rappels et réflexions," 174–79; Ellul, *Ethics of Freedom*, 385–95.
41. Ellul, *Ethics of Freedom*, 388.
42. Ellul, *Ethics of Freedom*, 385.
43. Ellul, *Ethics of Freedom*, 383.

5.1 Study the societal *maincurrents* below the surface.⁴⁴ Our society—including our politics and media as well as entertainment—is utterly superficial. Rather than obsessing about eye-catching headlines, tweets, and surface factoids, Christians should focus on the underlying, broader, basic "maincurrents" below the surface of our lives. That means focusing on things like the growth of technology, changes in jobs and work, longer-term trends and changes in family structure and sexual behavior, trends in education and information. This means reading more, longer, deeper studies, and more history—and fewer newspaper headlines. Don't watch television news with its breathless and absurd fifteen or twenty-second, "breaking news" stories and its addiction to celebrity gossip, and blood and guts spectacles.

5.2 Warn of *future consequences*, threats, and challenges.⁴⁵ Try to play the role of Ezekiel's "watchman on the wall" (33:1–20) watching for coming threats and opportunities, warning our world of what may be coming, thinking ahead about consequences of current activities and trends. "Presentism" is an enemy. Ellul rues the fact that most French intellectuals only became passionately interested in the Algerian war for independence from France in the late 1950s by which time violence and deep division was unavoidable. It was in the 1930s that the colonial situation was fluid and had a much better chance of being resolved relatively justly and peacefully but few paid attention. What else might be coming down the road that our world needs to hear about today?

5.3 Critique the *means*.⁴⁶ "The problem of means is radical and decisive in our society . . . characterized by the multiplication and power of our means and no longer by ends. . . . But it is always at the level of the means that the problems of humanity and morality are posed."⁴⁷ We must remind the political power that the means affect the character of the end. You can't achieve a just end if you use unjust means.

5.4 Address the "vast domain of the *psychic*"—the emotional, spiritual, and relational dimensions of life.⁴⁸ Our society full of fear and anxiety. People suffer the anguish of failure and loss and have few relationships to sustain them. Personal identity and self-image trouble many. Loneliness and isolation characterize many people's daily existence. The emotional and spiritual lives of many are in chaos. These are all areas where Christians could focus.

---

44. Ellul, *Hope*, 279–81. See also Ellul, *Presence*, 80.
45. Ellul, *False Presence*, 186–87.
46. Ellul, *Ethics of Freedom*, 402–8; Ellul, *Presence*, 39–62.
47. Ellul, "Theses," 193–94. My translation.
48. Ellul, *False Presence*, 185–86.

5.5 Advocate for the *truly poor*.[49] There are the "popular poor" in our world who are the subject of a vivid, guilty conscience and who have strong representation in the halls of political power. Of course their needs are real and it is good that their cause be advocated. But Ellul points out that there are others who we could call the "truly poor," the unpopular poor, the poor who are ignored and have no advocates. Kurds, Biafrans, and Tibetans are among those Ellul cited in the 1970s. Without doubt, it could be a special and distinctive focus of Christian political presence to be on constant lookout for those with no voice and come alongside as both servant and advocate.

6. Serve as ministers of reconciliation. Christians should also carry out a "ministry of reconciliation" in our alienated, partisan, segregated culture.[50] This should start within the church, working to concretely unite with other Christians, especially those of different backgrounds, identities, passions, challenges, and callings. Martin Luther King Jr. once observed that the most segregated hour of the week in America was Sunday morning at 11:00. What is still woefully absent in terms of racial diversity and unity is more and more true of political diversity. This is not just good for the church but it is a witness and example to the world around the church. They become the prototype of a unity-in-diversity for all to see. And they also bear witness to the relativity of politics: there is a unity that transcends our partisanship. They demonstrate this by the way they maintain and celebrate their commonality and unity even while retaining their political and personal distinctives. Only Christians can play this role, Ellul believes.[51]

The ministry of reconciliation means calming passion and debate, relativizing situations with humor, lowering the volume, and reducing tensions. It means helping opposing groups and individuals understand each other better and see each other's human value (while the political climate exacerbates difference and demonizes the other). Christians from opposite sides could serve as a communication channel among antagonistic groups where none may have existed or been thought possible. Reconciliation entails not just reaching out and bringing the other to our table but taking the time to speak to our own group on behalf of the other. Presumably we speak the language and have won the right to be heard by our group—so let's use that access to help our colleagues understand the fears, hopes, and foundational values and ideas driving the opposition. Denunciation and ignorance of the other are unlikely to help reconciliation.

7. Create a life outside of politics and the state. Ellul undoubtedly still shocks people inside and outside the church by statements such as this: "I

---

49. Ellul, *Betrayal*, 113–25.
50. Ellul, *False Presence*, 190–98.
51. Ellul, *Ethics of Freedom*, 379–80.

have to say that for most people today the involvement which is most useful and which best expresses Christian freedom, is involvement in anarchy."[52] Ellul sometimes sounds as though it is only the monolithic, totalitarian state of our era that justifies the extreme strategy of anarchism. But in other places he goes way back to Jesus, long before the rise of the modern technical-nation-state, to justify a rejection of the "archies" (the powers that oppress and control). Jesus, Ellul insists, shows a kind of disdain, a disregard for politics and the state.[53] In any case, Ellul's anarchy is not based on a belief in the possibility of a good society composed of naturally good human beings corrupted only by political repression. Rather, it is an extreme *strategic* position of resistance to a totalitarian state. It suggests to some of its adherents that a refusal to vote, pay taxes, or acknowledge any obligations or responsibilities to the established political or governing authorities—is the best political approach today.[54]

At the same time the state (or at least its status as authoritarian, totalitarian ruler of our lives) is rejected by our anarchy, we must create a life of freedom and meaning for ourselves as individuals—and as small, non-political communities.[55] We need a Yes alongside our No. This might include a "prevention club" for teenagers otherwise just hanging out on the street and prone to get into trouble. It could be a film club or a movement to protect the natural coastal environment from corporate development and political attack. It could be associations of various professionals (medicine, law, banking, the arts, etc.) gathering to reflect on their work with its challenges and possibilities. It could be creating community gardens in otherwise empty lots around town. The point is not just to refuse (or stop relying upon) the state's management of our lives but to step out beyond the state and politics and take positive creative initiative.[56]

52. Ellul, *Ethics of Freedom*, 395.

53. See Ellul, *Anarchy*. Ellul's version of anarchy resembles libertarianism on the political side, wishing to shrink the state. *Unlike* most libertarians, Ellul does not want to shrink the state only to allow *corporate* economic powers to crush individual and community freedom. Ellul wants to resist *all* the "arche" (powers, authorities, institutions that constrain human freedom). But neither is he a nihilist, simply destroying and resisting out of cynicism, meaninglessness, and a death wish. His version of anarchism is something like a "non-violent, individualistic populism."

54. Jerome Ellul recently sent me an opinion piece his grandfather drafted in 1986, "Je suis un abstentionniste" ("I am an abstainer" (from voting)), in reference to one of France's general elections of a new president. (We do not know if or where this piece might have been published). His reasons were that he didn't see any significant difference in the candidates and none of them really laid out any coherent plans that would actually address the real needs of the people. I don't think, however, that it was just one election; I think Ellul voted very rarely if at all.

55. Ellul, *Presence*, 96–99.

56. All of the preceding were projects of Ellul himself or his friends and students.

Reflections

Perhaps most *states* throughout history have been unjust, incompetent, and oppressive somewhere on a continuum from disappointing—to barely tolerable—to wicked, exploitive, and tyrannical. The transformation of virtually all modern states into technicized bureaucracies that grow almost unabated toward totalitarianism (soft or hard) is hardly disputable. Ellul also has good reason to view *politics* on a continuum from useless, ineffective, and illusory—to corrupt, power-hungry, divisive, and full of lies. We only have to look at politics and the state around the world today to see that we are in trouble almost everywhere.

I think Ellul is mostly right in his sociological-historical and theological perspectives on both politics and the state. With his Bordeaux colleague, Professor Jean-Louis Seurin, however, I think Ellul overstates the (negative) single essence of all modern states, perhaps allowing his theological view of an essentially diabolical, satanic state and political order to narrow his sociological-historical reading of the state and the political process.[57] There is, Seurin argues, a significant (even if not total) difference between a pluralistic constitutional democracy and a totalitarian dictatorship. A state that welcomes a naked will to power is different from one that institutionalizes checks and balances on power and corruption.

I would further argue that it is not immaterial whether the elected leader is a corrupt, uneducated, vulgar, narcissistic, television reality star like Donald Trump—or a well-educated, thoughtful, experienced, team-oriented, articulate leader like Barack Obama or Hilary Clinton (fully aware of their own imperfections in policy and practice!). Thus, I would feel irresponsible abstaining from study, debate, and voting, given such alternatives. The single-handed destruction Trump achieved in just one year in office is stunning.

I grant that much of politics is illusory and conducted in the world of images—while much of our reality is shaped in the bureaucracies of the beast, often yielding similar results whether under a Bush or a Clinton or even a Jintao, Putin, or Assad. But I maintain that there are differences, perhaps especially with regard to the havoc they can get us into. So I will continue to participate, though *without any illusions* that campaign promises can be completely fulfilled, ideals matched, or perfection achieved.

In an ironic twist, having accepted for decades now Ellul's argument that we must not allow measureable effectiveness to rule our decisions, I

For more of Ellul's story, in his own words, see the extended interviews in Troude-Chastenet, *Jacques Ellul on Politics*; and Ellul, *In Season*.

57. Seurin, "Jacques Ellul," 323–55.

am not at all deterred from voting by arguments that it doesn't matter, that it is not effective! Since when does an Ellulian make efffectiveness the decisive value? Same for my public and private arguments and debates: I do not imagine stopping just because I am not measurably effective. It's about truth and about faithful witness.

On the biblical theology side, I think the political picture is more complex and varied than Ellul has it. Not the least of my influences here is the deliberate choice of ordinary political vocabulary by Jesus and the early church to describe the social organization of their movement: e.g., *basilea, ecclesia, koinonia, dykaiosune, arête* . . . this is the vocabulary not just of ordinary politics but of the early church for its own life and reality. Ellul dismisses this usage but Jesus and the New Testament could have used a religious vocabulary of "temple," "synagogue," (or think of "mosque" and "ashram") to describe the new movement. But instead it deploys language and conceptuality that is mostly political. Christians are an *ecclesia* (political assembly of citizens, now translated (misleadingly) as "church"), a *basilea* (kingdom), and a *nation*. This is critical. The Christian movement is not *anti*-political; it is deliberately "*other*-political." Instead of accepting an identity as a full-on citizen exclusively or primarily of the USA, as a Christian I am both an alien pilgrim *and* an ambassador from the kingdom of God. What that ambassadorship looks like is pretty much what Ellul describes and I have summarized above.

In sum, I do not feel like giving up on political participation and reform as radically as Ellul seems to suggest in his reflections on anarchy. I do very strongly buy into Ellul's call for unity among Christians (or among Ellulians for that matter) across ideological and partisan camps. I agree with his call not just to be the chaplains and the "Amen corner" for existing political authorities and movements but to be their conscience. I love the call to bring something creative, redemptive, specific, unique, incommensurate with ordinary politics in a prophetic dialogue with power. I believe in a politics that is humble and based in neighborhoods and community groups, maybe extending to schools and other public services as well as local economies. Most of all I believe not just in cursing the darkness but in lighting a candle.

Bibliography

Ellul, Jacques. *Anarchy & Christianity*. Translated by Geoffrey W. Bromiley. Grand Rapids: Eerdmans, 1991.

———. *Apocalypse: The Book of Revelation*. Translated by George W. Schreiner. New York: Seabury, 1977.

———. *Betrayal of the West*. Translated by Matthew J. O'Connell. New York: Seabury, 1978.
———. *The Ethics of Freedom*. Translated by Geoffrey W. Bromiley. Grand Rapids: Eerdmans, 1976.
———. *False Presence of the Kingdom*. Translated by C. Edward Hopkin. New York: Seabury, 1972.
———. *Hope in Time of Abandonment*. Translated by C. Edward Hopkin. New York: Seabury, 1973.
———. *In Season, Out of Season: An Introduction to the Thought of Jacques Ellul. Based on Interviews with Madeleine Garrigou-Lagrange*. Translated by Lani K. Niles. San Francisco: Harper & Row, 1982.
———. *Living Faith*. Translated by Peter Heinegg. San Francisco: Harper & Row, 1983.
———. *The New Demons*. Translated by C. Edward Hopkin. New York: Seabury, 1975.
———. *The Political Illusion*. Translated by Konrad Kellen. New York: Knopf, 1967.
———. *The Politics of God, the Politics of Man*. Translated by Geoffrey W. Bromiley. Grand Rapids: Eerdmans, 1972.
———. "La politique modern lieu du demoniaque." In *La Politique*, edited by Patrick Troude-Chastenet, 161–88. Also, with slight revisions, in Ellul, *Living Faith*, 234–48.
———. *Presence in the Modern World*. Translated by Lisa Richmond. Eugene, OR: Wipf & Stock, 2016.
———. "Rappels et réflexions sur une théologie de l'état." In *Les Chrétiens et L'État*, by Jacques Jullien, Pierre L'Huillier, and Jacques Ellul, 129–82. Paris: Mame, 1973.
———. *The Subversion of Christianity*. Translated by Geoffrey W. Bromiley. Grand Rapids: Eerdmans, 1986.
———. *The Technological Society*. Translated by John Wilkinson. New York: Vintage, 1964.
———. "Theses sur foi chrétienne et politique." In *La Politique*, edited by Troude-Chastenet, 189–99. Bordeaux-Le Bouscat, France: L'Esprit du Temps, 2008.
Holloway, James, ed. *Introducing Jacques Ellul*. Grand Rapids: Eerdmans, 1970.
Seurin, Jean-Louis. "Jacques Ellul: L'interpretation de la politique à la lumière de la Bible." In *Sur Jacques Ellul: Un Penseur de Notre Temps*, edited by Patrick Troude-Chastenet, 323–55. Bordeaux-Le Bouscat, France: L'Esprit du Temps, 1994.
Troude-Chastenet, Patrick. "Jacques Ellul et la politique." In *La Politique*, edited by Patrick Troude-Chastenet, 33–59. Bordeaux-Le Bouscat, France: L'Esprit du Temps, 2008. Also published in *Jacques Ellul: Penseur Sans Frontières*, edited by Patrick Troude-Chastenet, 123–47. Bordeaux-Le Bouscat, France: L'Esprit du Temps, 2005.
———. *Jacques Ellul on Politics, Technology, and Christianity: Conversations with Patrick Troude-Chastenet*. 1994. Translated by Joan Mendès France. Eugene, OR: Wipf & Stock, 2005.
———, ed. *La Politique (Cahiers Jacques Ellul No. 5)*. Bordeaux-Le Bouscat, France: L'Esprit du Temps, 2008.
———, ed. *Sur Jacques Ellul: Un Penseur de Notre Temps*. Bordeaux-Le Bouscat, France: L'Esprit du Temps, 1994.
Walzer, Michael. *In God's Shadow: Politics in the Hebrew Bible*. New Haven: Yale University Press, 2012.

5

Jacques Ellul and the Nonviolent Movements of His Time in France[1]

BY Frédéric Rognon

> *Frédéric Rognon is Professor of Philosophy at the Faculty of Protestant Theology at the University of Strasbourg (France), President of the Justice and Prison Chaplaincy Commission of the Fédération Protestante de France, and Editor of the journal* Foi & Vie. *Among his published works are* Jacques Ellul: Une pensée en dialogue *(Geneva: Labor et Fides, 2007, 2013) and* Générations Ellul: Soixante héritiers de la pensée de Jacques Ellul *(Geneva: Labor et Fides, 2012).*

* * * *

IN JACQUES ELLUL'S TIME, three major movements—clearly distinct, occasionally rival, often allied—occupied the scene of nonviolent thought and action in France. The oldest of the three movements was the Mouvement International de la Réconciliation. The MIR was founded in 1923 as the French branch of the International Fellowship of Reconciliation (IFOR), which had been established in 1919. It is a movement of Protestant origin with a spiritual orientation, dedicated to promoting "evangelical nonviolence."[2] The *MIR* is a member of the Fédération Protestante de France.[3] During the Sixties and the Seventies, when Jacques Ellul was

---

1. Translated from the French by Jason Dean.

2. Translator's note [henceforth TR]: In this article the adjective "evangelical" refers to the Gospel (Greek, *Evangelium*) of Christ, not the fundamentalist religious movement born in the United States at the beginning of the twentieth century.

3. TR: The FPF regroups most French Protestant Churches, including the largest, the Église Réformée de France (ERF) [Reformed Church of France], Ellul's Church, which merged with the Église Évangélique Luthérienne de France (EELF) [Evangelical

active in the debate on nonviolence, the main leaders of the MIR were Henri Roser (1899–1981), André Trocmé (1901–1971), Jean Lasserre (1908–1983), and Jean Goss (1912–1991).

The second nonviolent movement was associated with the Communautés de l'Arche [Communities of the Ark], founded in 1948 by the Catholic philosopher, artist, poet, and musician Lanza del Vasto. After meeting Gandhi in India in 1937, Lanza del Vasto created rural, agricultural, and artisanal communities of work, prayer, and nonviolent action modeled on Gandhi's ashrams. The Communautés de l'Arche are therefore a communal and interfaith movement of Gandhian sensibility. Among its most important organizers were Lanza del Vasto (1901–1981), Pierre Parodi (1923–1989), and Jo Pyronnet (1927–2010).

Founded in the wake of the May 1968 uprisings in Paris, the Mouvement pour une Alternative Non-Violente (MAN), the most recent of the three nonviolent movements, adopted a secularist and decidedly political perspective. Its ideology is akin to self-managing socialism; its critique of capitalism often borrows from the Marxist lexicon; its assessment of statist socialism is avowedly anarchist. The founding members and principal leaders of the MAN are Jean-Marie Muller (b. 1939), Christian Mellon (b. 1943), Christian Delorme (b. 1950), and Jacques Semelin (b. 1951). All four are Catholics, two of them (Mellon and Delorme) priests. They are, nevertheless, committed to a non-confessional approach to nonviolence.

Together, these three movements represent a few hundred activists and a few thousand sympathizers. Ellul had real affinities and profound disagreements with each of these movements. The exchanges of views he had with them were frank, even blunt. We will examine the points of agreement and disagreement between the parties in order to clarify the particular characteristics of the Ellulian approach to nonviolence.

Affinities

The principal affinities between Jacques Ellul and the three French nonviolent movements can be traced back to their analyses of the phenomenon of violence. In his 1972 book *Violence: Reflections from a Christian Perspective*, Ellul enumerated five laws of violence.[4] These five laws include: (1) Continuity: the use of violence triggers a spiral of violence, "once a man

---

Lutheran Church of France] to form the Église Protestante Unie de France (EPUF) [United Protestant Church of France] in 2012.

4. Ellul, *Violence*, 93–108. For the French text, see Ellul, "Contre les violents," 501–639.

has begun to use violence he will never stop using it."⁵ (2) Reciprocity: the use of violence invariably provokes a violent response; it is therefore pointless to ask who initiated it. (3) Sameness: violence is violence, there is no such thing as just and liberating violence as opposed to unjust and enslaving violence. (4) Self-generation: the recourse to violence produces a violent society; consequently "the end does not justify the means.... On the contrary, evil means corrupt good ends."⁶ (5) Continual justification: "the man who uses violence always tries to justify both it and himself," because "the plain fact is that violence is never 'pure.'"⁷ The three French nonviolent movements of the sixties and seventies concur unhesitatingly with these five characteristics of violence, which Ellul calls "laws," in order to better disqualify the use of violence.

In his condemnation of violence, Ellul further criticizes the criteria used to define "just war," demonstrating their inadequacy.⁸ The criterion of "lesser evil" proves to be particularly equivocal, since one can never know what the point of comparison is. The criterion of "last resort" authorizes all sorts of abuses. And the injunction to "kill without hate" is completely unrealistic.⁹ Here, too, there is agreement between Ellul and the nonviolent movements. They agree that there is no such thing as a "just war," and that all justifications of violence, especially Christian justifications, must be rejected.

Divisions nevertheless appeared in this common foundation, and Jacques Ellul differentiated himself by criticizing the three movements. We will examine Ellul's relationship with each of these movements, beginning with the most recent one, and the one with which the divide appears

---

5. Ellul, *Violence*, 94.

6. Ellul, *Violence*, 102. Cf. concerning the fourth law: Ellul, *Critique of the New Commonplaces*, 294–303. In this criticism of the commonplace that "the ends justify the means," Ellul notes that "the nobler, loftier, and more just the ends that man assigns himself, the more monstrous and inhuman the means he will employ" (296); "vile instruments, destined to disappear once the end has been realized" (296). Instead, in reality, "*the use of violence leads inevitably to the establishment of a dictatorship and to the denial of freedom*" (300, underlined by Ellul in original). And again, "a law established by violence will always be injustice.... A perfect Society organized by shedding blood, even that of guilty men, will always be a forced labor camp" (303). For the French edition, see Ellul, *Exégèse des nouveaux lieux communs* (Paris, La Table Ronde, 2004).

7. Ellul, *Violence*, 103–4.

8. Ellul, *Violence*, 5–8.

9. Cf., on "just war": Ellul, "En toute liberté" ["Violence and nonviolence," untranslated in English], 3. In this text, Ellul writes that Thomas Aquinas must be "joking" when he defends the criterion of "lesser evil," since "only God can take that measure." See also Ellul and Troude-Chastenet, *Jacques Ellul on Politics*, 39. For the French text, see Ellul and Troude-Chastenet, *À contre-courant. Entretiens*, 94–95.

the deepest, the Mouvement pour une Alternative Non-Violente (MAN). We will then work backward in time toward the Communautés de l'Arche and, finally, to the Mouvement International de la Réconciliation (MIR), to which, at first blush, Jacques Ellul seems closest. But, as we will see, that is only an appearance.

Jacques Ellul and the Mouvement pour une Alternative Non-Violente (MAN)

Jacques Ellul published two articles in the journal of the MAN, *Alternatives Non-Violentes*.[10] The first was a commissioned article. The professor from Bordeaux was one of fifteen personalities, including Henri Roser of the MIR and Pierre Parodi of the Communautés de l'Arche, invited in 1977 to react to the MAN's *Texte d'Orientation Politique (TOP)* [*Declaration of Political Orientation*]. This document of 130 pages was a manifesto for the fledgling movement, published the preceding year under the title "For Self-Managing Socialism: A Political Nonviolence."[11] The authors of the *Declaration* develop a critique of capitalism, which they see as a "society of injustice and violence," but also of statist socialism with its totalitarian orientation.[12] They formulate a certain number of propositions for a strategy of nonviolent transition to self-managing socialism and for the creation of a nonviolent popular defense force in France. They insist on the efficacy of nonviolence when strategy and goals coincide, affirming that "nonviolence offers effectual tools for attaining the goal of a society based on self-managing socialism."[13] The following statement from the *Declaration* summarizes their position within the larger nonviolent movement:

> All too often, proponents of nonviolence confine themselves to a moralistic perspective, preaching the "spiritual conversion" of the individual while ignoring the necessity of promoting a political revolution, and imagining that the revolution will only come about as a result of the sum of the changes in the lives of all individuals. Such a conception of nonviolence leads to the condemnation of history as a history of violence and the desire to be above the fray, whereas, on the contrary, it is necessary to

10. Ellul, "Une non-violence privée de son fondement" ["Nonviolence Stripped of its Foundation," untranslated in English], 15–18; Ellul, "Violence et terrorisme" ["Violence and Terrorism," untranslated in English], 3–8.
11. "Pour le socialisme autogestionnaire," 1–130 [untranslated in English].
12. "Pour le socialisme autogestionnaire," 7.
13. "Pour le socialisme autogestionnaire," 4, cover.

be present at the heart of historical events and to demonstrate the efficacy of nonviolent methods of action. We therefore reject all moralistic approaches to nonviolence that ignore that man is a product of society and that, in order to transform himself, though he must act by himself, he can only achieve his goal if he acts with others to transform society. Of course, the personal dimension should not be scorned, but it must remain in relation to the collective dimension. It is by acting together that we can break the chains that enslave us. The working class carries the memories of its wounds and its dead; it therefore cannot accept the moralizing words of bourgeoisie preaching, a form of nonviolence predicated upon loving enemy that becomes a form of class collaboration.[14]

Responding to the *Declaration*, Jacques Ellul first notes points of agreement concerning the analysis of violence and society, while regretting that the Marxist orientation of the document neglects the technological changes that characterize the twentieth century and the entry into the "technological system." But Ellul's primary criticisms concern the anthropological underpinnings of the project, its focus on political strategies, and its disregard of the spiritual conditions of nonviolence. Hence the title of Ellul's contribution: "Nonviolence Stripped of its Foundation." Ellul writes:

> This is the central weakness of the project: it depends entirely on the conviction that man is fundamentally good and that it suffices for man to be placed in good structures for his goodness to be revealed. . . . This, then, is the core of my disagreement. The authors of this project are violently critical of the moral and religious aspects of nonviolence. They strive to disconnect nonviolence from this foundation in order to pursue political analysis, a political project purported to be realistic, feasible, efficacious. . . . [Now] all of this presupposes a change in the personality of each one of us, a moral and spiritual conversion.[15]

Here, Ellul stigmatizes the idealism of the framers of the *Declaration*, who overestimate the virtuous consciousness of men and women while neglecting the necessity of individual efforts to enter into nonviolence. He notes the irony of the references to Gandhi and Martin Luther King, Jr., who were first and foremost men of prayer. Last, he takes up the argument about efficacy (a signature expression of the technological mentality) and uses it

---

14. "Pour le socialisme autogestionnaire," 3–4.
15. Ellul, "Une non-violence privée de son fondement," 17.

against its authors, arguing that "deprived of its spiritual foundation, nonviolence is necessarily less efficacious . . . than violence."[16]

One of the leaders of the MAN, Jean-Marie Muller, objected strenuously to Ellul's criticism,[17] denouncing it as too superficial, and drawing a distinction between legitimate moral approaches to nonviolence and approaches that are only moralistic or moralizing while rejecting the precondition of moral and spiritual conversion to nonviolent action, affirming that "it is through action that we are converted."[18] Jean-Marie Muller accuses Ellul of using the language of Christianity—"late great Christianity," he calls it—in order to obfuscate the legitimate secularization of politics, and rebukes him for the scepticism which leads Ellul to discount the possibility of breaking the fatality of violence in history.[19]

> Jacques Ellul denies nonviolent action the possibility of being efficacious in history. His nonviolence is a prisoner of his moral and spiritual vision. It can only be a protest and a witness, it cannot become an action. That is the heart of our disagreement. . . . We do not wish to counter Jacques Ellul's scepticism with triumphalism. We do however have the audacity to challenge him with our hope and our determination.[20]

Thus, the dialogue between Jacques Ellul and the MAN became a dialogue of the deaf. The confusion between moral and moralistic approaches, for which Ellul bears responsibility, is doubled by the stereotype, equally unjust, of a Jacques Ellul devoid of hope for the future. This 1977 divorce did not prevent Ellul from publishing another article in the MAN journal five years later, this time on the subject of terrorism.[21] In this paper, Ellul identifies the roots of terrorism in the state, a position that would appeal to readers' libertarian sensibilities. But he goes on to link terrorism to politics in general. The convergence between Ellul and the MAN reappears however when the professor from Bordeaux remarks that the origin of terrorism is to be found in the separation of politics from

---

16. Ellul, "Une non-violence privée de son fondement," 17. Five years, later expounding on this text, Jacques Ellul described Lech Walesa, the leader of *Solidarnosc*, as an example of a man of faith and nonviolence with, at once, flexibility and firmness, and "the model of the politically engaged Christian." Ellul, "Lech Walesa et le rôle du christianisme," 40–47 (n. 42).

17. Muller, "Tribune libre," 61–66.

18. Muller, "Tribune libre," 63.

19. Muller, "Tribune libre," 64.

20. Muller, "Tribune libre," 66.

21. Ellul, "Violence et terrorisme."

morals, leading to a situation where all means to gain political power are legitimate.[22] Thus, a consensus is reached on the basis of the principle of consistency between means and ends.

Jacques Ellul and the Communautés de l'Arche

In my contribution to the 2012 Bordeaux conference on the thought of Jacques Ellul, I analyzed his relations with Lanza del Vasto.[23] My analysis showed that these two thinkers rarely quoted each other, and when they did they criticized each other in no uncertain terms. The first grievance concerns their respective ways of life. In his book *De la révolution aux révoltes* (1972), Ellul cites the Communautés de l'Arche as a typical example of a counter culture devoured by the technological system: "It is not sufficient to proclaim that in these counter-societies the authority of the state is repudiated. In my room, I can repudiate private property and imagine that I have annihilated the power of trusts. I can even create an ashram . . . but Lanza del Vasto has not changed the economic system, and these 'anarchist' groups have not modified in the least modern statism."[24] What Ellul has identified here is the inability of the agricultural and artisanal communities founded by Lanza del Vasto to constitute a credible alternative to the technological system. In a similar vein, Jacques Ellul stigmatizes, in *Violence*, the pacifist idealism of those who, like the hippies, live off the fat of a technological society of which they are "a product of the luxury," and to which they contribute by providing the creative and recreational elements it lacks.[25]

Inversely, Lanza del Vasto blamed Ellul for merely condemning urban and inhumane technological civilization while fully participating in it, rather than proposing a communal alternative, as he himself had done.[26] Thus, the grievances between the two men appear as mirror images: the inanity of the ashrams answers the vain and futile inconsistency of the intellectuals.

There are, however, other points of divergence between the Ellulian approach to nonviolence and that developed by the Communautés de

22. Ellul, "Violence et terrorisme," 7. Jacques Ellul devoted another text to the phenomenon of terrorism, emphasizing the importance of psychological action, and moving toward a description of the mechanisms of intellectual terrorism and the relations of submission within businesses in democratic societies: see Ellul, "Terrorisme et violence psychologique" ["Terrorism and Psychological Violence"], 43–61.

23. Rognon, "Jacques Ellul et Lanza del Vasto," 211–26.

24. Ellul, *De la révolution aux révoltes* [*From Revolution to Revolts*], 324–25n120.

25. Ellul, *Violence*, 119–22.

26. Dupont, "L'œuvre politique de Lanza del Vasto" ["The Political Work of Lanza del Vasto," untranslated in English], 431.

l'Arche. The first point concerns the necessity of violence. This question was debated at a round table discussion held on April 13, 1967, at the University of Bordeaux in which Ellul took part, along with Jean Lasserre, of the MIR, and Jo Pyronnet of the Communautés de l'Arche.[27] On this occasion, Ellul argued the thesis of a necessary though illegitimate violence: "Violence is a necessity that nothing can justify."[28] He went on to say: "I have therefore admitted violence as a necessity. But for me, necessity is neither reason nor virtue. We should not boast of being bound by necessity, any more than we boast of being bound and tied and falling by the force of the law of gravity."[29] In his book *A Critique of the New Commonplaces*, Ellul had already stigmatized the saying "making a virtue of necessity," outside of the context of violence.[30] At the round table discussion in Bordeaux, Jo Pyronnet affirmed that "it is possible to construct a human community from which violence is excluded,"[31] citing as an example his experience at the Communautés de l'Arche, whose structures are conceived and developed in such a way that profit, domination and property are eliminated and replaced by a spirit of service and sharing. For Ellul, such a small-scale communal laboratory of nonviolence in no way disproved his thesis of violence as a necessary part of all social life.[32]

The last point of disagreement between Ellul and the members of the Communautés de l'Arche concerns the choice of the word "violence" to refer to spiritual combat. In *Violence*, Ellul defends the only (in his eyes) legitimate violence: the one he calls "the violence of love," i.e., the combat consisting of overcoming evil with good, without giving in to the temptation of physical or psychological violence.[33] Lanza del Vasto could in no way condone the use of the term "violence" to designate an undertaking that integrally respected the dignity and integrity of every human being.

---

27. Ellul, Abribat, Day, Lasserre, and Pyronnet, "Table ronde," 101–26.

28. Ellul, Abribat, Day, Lasserre, and Pyronnet, "Table ronde," 101.

29. Ellul, Abribat, Day, Lasserre, and Pyronnet, "Table ronde," 115. Ellul revisited the idea of inevitable violence, stemming from necessity, but not legitimate, in Ellul, *Violence*, 84–93, 127–45 (particularly 139–40).

30. Ellul, *Critique of the New Commonplaces*, 171–78.

31. Ellul, Abribat, Day, Lasserre, and Pyronnet, "Table ronde," 111.

32. In "Les fondements éthiques de la non-violence" ["The Ethical Foundations of Nonviolence"], 260–79 [269], Jean-Marie Muller, the cofounder of MAN, takes a position very close to that of Jacques Ellul when he writes that "justifying violence by necessity is the surest way to make violence necessary," adding, "if I am trapped by necessity into using violence against my adversary, I should have the courage to not justify myself with excuses."

33. Ellul, *Violence*, 160–75.

Concerned with clearly distinguishing violence and brutality, on the one hand, and force and struggle, on the other, in order to create the possibility of nonviolent combat, the disciple of Gandhi abhorred all semantic and conceptual confusion on this point: how is it possible to go from physical and psychological violence to the exaltation of spiritual violence, all the while conflating the two fundamentally different phenomena under the heading of "violence"?[34] In the eyes of Lanza del Vasto, what Ellul called the "violence of love" was, in reality, less a matter of "violence" (the negation of the dignity and the integrity of the person) than of "spiritual combat" (fighting with the weapons of the Holy Spirit, of which the apostle Paul speaks in his letter to the Ephesians).[35] These two realities must not be confused at any cost, since they are mutually exclusive. In fact, Ellul makes the same point. "Spiritual violence and the violence of love totally exclude physical or psychological violence. Here the violence is that of the intervention of the Spirit of God."[36] Thus, on this point, the disagreement between Jacques Ellul and the Communautés de l'Arche is essentially terminological.

Jacques Ellul and the Mouvement International de la Réconciliation (MIR)

Of the three French nonviolent movements contemporary to his life, the MIR is indisputably the one with which Ellul had the strongest affinities. To be convinced of this one only needs to look at the four prominently situated references to Jean Lasserre, leader of the MIR, in *Violence*, including a laudatory recommendation and a long, heartily approved citation.[37] Another indication is the fact that *Cahiers de la Réconciliation*, the official publication of the MIR, opened its columns five times to Jacques Ellul: (1) as early as May 1939 for one of its first articles on the risk of dictatorship in France in response to recent events;[38] (2) again in 1972 for a reflection on the relationship between nonviolence and the law, in which Ellul defended the right to conscientious objection;[39] (3) this same issue contained an interview with Ellul on the occasion of the publication of the original

34. Lanza del Vasto, *Technique de la non-violence*, 28–30.
35. Eph 6:10–20.
36. Ellul, *Violence*, 169.
37. Ellul, *Violence*, 14, 16, 17, and 102. The laudatory recommendation on p. 16 finds Ellul noting that Pastor Jean Lasserre is "one of France's noblest champions of nonviolence." The long citation is on 102–3.
38. Ellul, "Péril à l'intérieur" ["Peril Inside"], 9–11.
39. Ellul, "La non-violence et la loi" ["Non-violence and Law"], 3–6.

French edition of *Violence* (in which the author makes some additional remarks and members of the MIR manifest their approval);[40] (4) in 1981 *Cahiers* reprinted an article which had first appeared in *Réforme*, praising such peacemakers as Gandhi, Martin Luther King, Jr, Anwar Sadat, and Lech Walesa;[41] and (5) finally, in 1984, *Cahiers* reprinted an article from *Sur le Roc* on the nuclear arms freeze.[42]

In this latter paper, Ellul takes a position that would bring him into disagreement with the MIR for the first time. The Fédération Protestante de France (FPF), in response to a campaign led by the MIR and contrary to the opinion of the neo-Calvinist theological seminary in Aix-en-Provence, pronounced itself in favor of a nuclear arms freeze. As a Christian, Ellul came out in favor the freeze but opposed the churches making demands on the government: "If I was President, Ellul said, I would have no choice but to continue the arms program, including the nuclear arms program."[43] That is why "the Christian, as a Christian, can assert his position of nonviolence, witness to it, take action to diminish the risks of conflict, and attempt to influence opinion on the question of abandoning nuclear arms but can absolutely not ask the government to implement such a policy."[44] At the risk of alienating his friends and supporters among the activists for evangelical nonviolence, Ellul once again places the mission of Christian nonviolence outside of the political arena, far from the illusion of politics.[45]

But the principal stumbling block between Ellul and the MIR was the tension that the professor from Bordeaux introduced between "nonviolence" and "nonpower."[46] In effect, Ellul establishes a three-term dialectic between

40. Monod et al., "Questions à Jacques Ellul à propos de . . . son dernier ouvrage: *Contre les violents*" ["Questions for Jacques Ellul on the publication of his most recent work: *Violence*"], 7–12.

41. Ellul, "Non-violence quand même" ["Nonviolence nevertheless"], 16. In this article, Ellul offers the following defense of nonviolence: "It is the only way to preserve the honor of mankind. It is not triumphal efficacy, it is the truth. Only though nonviolence can mankind have a history" (22). These assertions appear to be a response both to the exaltation of the criterion of efficacy by MAN activists and to the disqualification of nonviolence as inefficacious, and therefore ahistoric, by Paul Ricœur (cf. Paul Ricœur, "L'homme non-violent et sa présence à l'histoire," 317–35; Rognon, "Conflits, paix et non-violence dans *Histoire et vérité*," 111–22).

42. Ellul, "Le débat nucléaire" ["The nuclear Debate"], 18–19.

43. Ellul, "Le débat nucléaire" ["The nuclear Debate"], 18.

44. Ellul, "Le débat nucléaire" ["The nuclear Debate"], 18.

45. Ellul, *Political Illusion*. For the French text, see Ellul, *L'illusion politique*.

46. Ellul, "Ethics of Nonpower," 204–12; Ellul, "Power of Technique," 242–47; Ellul, *Subversion of Christianity*, 164–66; for the French text, see Ellul, *La subversion du christianisme*; Ellul, *What I Believe*, 149–51; for the French text, see Ellul, *Ce que je crois*; Ellul, *If You Are the Son of God*, 84–86; for the French text, see *Si tu es le fils de*

power, powerlessness and nonpower. Power is the capacity to do things; powerlessness is the incapacity to do them; and nonpower, the choice not to do things that one has the capacity to do, renouncing that which one is capable of doing, "*even . . . in order to defend [one's] life.*"[47] Nonpower, therefore, has nothing to do with powerlessness. Speaking to Christians, Ellul says that Jesus, the all-powerful, adopted an attitude of nonpower on multiple occasions, and not just nonviolence—by asking John the Baptist to baptize him, by resisting the three temptations of Satan to reveal his power, by declining to call on armies of angels during his arrest, and by refusing the possibility of climbing down from his cross.[48] Ellul sees in each of these examples an application of Jesus's teaching in the Sermon on the Mount. For him there is no such thing as "self-defense."[49] Following in the steps of Christ, Christians should adopt the path of nonpower.

As these examples show, Ellul invokes the principle of nonpower when he enumerates the diverse facets of what, in the Gospel, is most contrary to human nature. The historic departure from this stance is a critical part of what Ellul calls the "subversion of Christianity" over the centuries.[50] "And what about another concept that seems to be essential in the life of Jesus Christ, that of weakness [nonpower], which is linked with antipolitics? What can be more the opposite of what we are?"[51] In effect,

> truly the spirit of power lies deep in the human heart. How truly intolerable, then, is a message, and even more so a life, that centers on weakness [nonpower]. Not sacrifice on behalf of a cause that one wants to bring to success, but in all truth love for nothing, faith for nothing, giving for nothing, service for nothing. Putting others above oneself. In all things seeking the interest of others. When dragged before courts, not attempting any defense but leaving it to the Holy Spirit. The renunciation of power is infinitely broader and harder than nonviolence (which

---

Dieu. Souffrances et tentations de Jésus, in *Le défi et le nouveau, Œuvres théologiques 1948–1991*, 937–1016; Ellul and Troude-Chastenet, *Jacques Ellul on Politics*, 28; Ellul, *Théologie et Technique* (untranslated in English) , 64–67, 179–234 (particularly 221–22), 261–96, 307–52 (particularly 310–28).

47. Ellul, *If You Are the Son of God*, 84.

48. Ellul, *What I Believe*, 149–51. These biblical accounts can be found in Matt 3:13–17; 4:1–11; 26:52–53; 27:39–44; Mark 1:9–11; 15:29–32; Luke 3:21–22; 4:1–13; 23:35–37; John 1:19–34; and 18:11. See also Phil 2:5–11.

49. Ellul, *If You Are the Son of God*, 84.

50. Ellul, *Subversion of Christianity*, 154.

51. Ellul, *Subversion of Christianity*, 164–65.

it includes). For nonviolence allows of a social theory, and in general it has an objective. The same is not true of nonpower.[52]

The relationship between nonpower and nonviolence can be represented as concentric circles, since nonviolence is "contained" within nonpower. And yet, the choice of nonpower "does not rule out occasional acts of violence."[53] It is in *Théologie et Technique* that Jacques Ellul expressed his position on this subject most clearly:

> it is evident that [nonviolence] is englobed in [nonpower], but is not identical to it. In effect, nonviolence has become a strategy or a tactic for winning a political combat. Here again, we see the triumph of the technological spirit: the proponents of nonviolence are called upon to justify themselves by proving that nonviolence is efficacious and can meet with success. This is perhaps legitimate on a political level, but one must be aware that it signifies adhering to the technological system. It means recognizing that that which is not efficacious is worthless! That is the weakness of this position: if, in order to attain a given result, nonviolence is useful, and one uses it, then one has entered into violence. And this concern with efficacy is unvaryingly the cause of crises within nonviolent movements, and just as inevitably the proponents of nonviolence are overrun by a violent wing.
>
> If one seeks to prove that, thanks to Gandhi, one can defeat the enemy, then nonviolence becomes a technique among others (though a gentle one!), which must comply with the technological spirit of success and efficacy. This amounts to admitting that one cannot convince the members of society other than by proof of efficiency, and that saying "Of course, it's not efficacious (so what?), but it is good, just and true" has no value whatsoever! It amounts to entering the technological system whose spirit of power constitutes the norm and the criteria recognized by all. . . . Thus, nonpower undercuts all these misunderstandings. True, it is not efficacious, and that is precisely why it is the only critical alternative to the technological system. All the rest is anecdotal.[54]

The distinction between nonviolence and nonpower was never brought up for debate by the members of the MIR. It is as though the Ellulian position was too radical to be admitted by the movement, despite

---

52. Ellul, *Subversion of Christianity*, 166.
53. Ellul, *What I Believe*, 149.
54. Ellul, *What I Believe*, 318–19.

its similarities with some of the intuitions of Jean Lasserre and his friends, who were more concerned with being faithful to Christ than with being efficacious at any cost.[55] No doubt some of them could have (or could still?) endorse the Ellulian thesis according to which "today only a nonuse of power has a chance of saving the world."[56]

Conclusion

We can summarize the specificity of Jacques Ellul's approach to violence and nonviolence in this way: (1) deconstruction of the justifications of violence, but recognition of a necessary violence and even of a legitimate violence (the violence of love); (2) affirmation of the spiritual foundation of nonviolence, but criticism of a technological nonviolence obsessed with the goal of efficiency; (3) promotion of the concept of "nonpower," which alone is faithful to the biblical roots of true nonviolence. Navigating between affinities and dissimilarities, misunderstandings and mutual stereotyping, Jacques Ellul was able, in this field of analysis as in many others, to trace his own route through the many currents of thought and action of his time.

Bibliography

Del Vasto, Lanza. *Technique de la non-violence*. Paris: Denoël, 1971.
Dupont, Bernard. *L'œuvre politique de Lanza del Vasto*. PhD thesis, Université de Nancy II, Faculté de droit et de sciences économiques, February 1977.
Ellul, Jacques. *Ce que je crois*. Paris: Grasset, 1987.
———. "Contre les violents." In *Le défi et le nouveau. Oeuvres théologiques 1948–1991*, by Jacques Ellul, 501–639. Paris: La Table Ronde, 2007.
———. *A Critique of the New Commonplaces*. Translated by Helen Weaver. Eugene, OR: Wipf & Stock, 2012.
———. *De la révolution aux révoltes*. Paris: La Table Ronde (La petite Vermillon n°345), 2011.
———. "Le débat nucléaire." *Sur le Roc* (April 1984) 8. Republished in *Cahiers de la Réconciliation*, 51st year, n°11 (November 1984) 18–19.
———. "En toute liberté: violence et non-violence." *Réforme* n°1383–1384 (August 1973) 3.
———. "The Ethics of Nonpower." In *Ethics in an Age of Pervasive Technology*, edited by Melvin Kranzberg, 204–12. Boulder, CO: Westview, 1980.

---

55. Lasserre, *La guerre et l'Évangile*, 24–28; English translation: Lasserre, *War and the Gospel*, 18–20. The author subordinates efficiency to faithfulness, refuses to consider efficiency a moral criterion and protests against the corruption of Christianity by the cult of efficiency.

56. Ellul, *What I Believe*, 151.

———. *If You Are the Son of God: The Suffering and Temptations of Jesus*. Translated by Anne-Marie Andreasson Hogg. Eugene, OR: Wipf & Stock, 2014.

———. *L'illusion politique*. Paris: La Table Ronde, 2004.

———. "Lech Walesa et le rôle du christianisme." *Esprit* (March 1982) 40–47.

———. "La non-violence et la loi." *Cahiers de la Réconciliation* n°7–8 (July–August 1972) 3–6.

———. "Une non-violence privée de son fondement." *Alternatives Non-Violentes* n°20–21 (January 1977) 15–18.

———. "Non-violence quand même." *Réforme* n°1904 (October 1981) 16. Republished in *Cahiers de la Réconciliation* n°12 (December 1981) 22–23.

———. "Péril à l'intérieur." *Cahiers de la Réconciliation* n°5 (May 1939) 9–11.

———. *The Political Illusion*. Translated by Konrad Kellen. Eugene, OR: Wipf & Stock, 2015.

———. "The Power of Technique and the Ethics of Non-Power." In *The Myths of Information: Technology and Postindustrial Culture*, edited by Kathleen Woodward, 242–47. Madison, WI: Coda, 1980.

———. *La subversion du christianisme*. Paris: La Table Ronde, 2001.

———. *The Subversion of Christianity*. Translated by Geoffrey W. Bromiley. Eugene, OR: Wipf & Stock, 2011.

———. "Terrorisme et violence psychologique." In *La violence dans le monde actuel*, by Centre d'études de la civilisation contemporaine, 43–61. Paris: Desclée de Brouwer, 1968.

———. *Théologie et Technique: Pour une éthique de la non-puissance*. Edited by Yves Ellul and Frédéric Rognon. Geneva: Labor et Fides, 2014.

———. *Violence: Reflections from a Christian Perspective*. Translated by Cecelia Gaul Kings. Eugene, OR: Wipf & Stock, 2011.

———. "Violence et terrorisme." *Alternatives Non-Violentes* n°45 (Summer 1982) 3–8.

———. *What I Believe*. Translated by Geoffrey W. Bromiley. Grand Rapids: Eerdmans, 1989.

Ellul, Jacques, J.-P. Abribat, Hem Day, Jean Lasserre, and Joseph Pyronnet. "Table ronde: Violence, non-violence et revolution." In *Violence humaine, violence libératrice?*, by Roland Barois et al., 101–26. Paris: Le Centurion (Approches), 1968.

Ellul, Jacques, and Patrick Troude-Chastenet. *À contre-courant. Entretiens*. Paris: La Table Ronde, 2014.

———. *Jacques Ellul on Politics, Technology, and Christianity: Conversations with Patrick Troude-Chastenet*. Translated by Joan Mendès France. Eugene, OR: Wipf & Stock, 2005.

Lasserre, Jean. *La guerre et l'Évangile*. Paris: La Réconciliation, 1953.

———. *War and the Gospel*. London: James Clarke & Co Limited, 1962.

Monod, Ambroise, et al. "Questions à Jacques Ellul à propos de . . . son dernier ouvrage: *Contre les violents*." *Cahiers de la Réconciliation* n°7–8 (July–August 1972) 7–12.

Muller, Jean-Marie. "Les fondements éthiques de la non-violence." *Dire la Guerre, penser la Paix. Actes du Colloque international de Strasbourg, 14–16 mai 2012*, directed by Frédéric Rognon, 260–79. Geneva: Labor et Fides (Le Champ éthique n°62), 2014.

———. "Tribune libre: Réponse à Jacques Ellul." *Alternatives Non-Violentes* n°24–25 (August–October 1977) 61–66.

"Pour le socialisme autogestionnaire: une non-violence politique." *Alternatives Non-Violentes* n°15-16 (April 1976) 1-130.

Ricœur, Paul. "L'homme non-violent et sa présence à l'histoire." In *Histoire et vérité*, by Paul Ricœur, 317-35. Paris: Éditions du Seuil, 1951.

Rognon, Frédéric. "Conflits, paix et non-violence dans *Histoire et vérité*." In *La jeunesse d'une pensée. Paul Ricœur à l'Université de Strasbourg (1948-1956)*, directed by Daniel Frey, 111-22. Strasbourg: Presses Universitaires de Strasbourg, 2015.

———. "Jacques Ellul et Lanza del Vasto. Regards croisés sur deux témoins de l'espérance au cœur d'un monde sans issue." In *Comment peut-on (encore) être ellulien au XXIe siècle ? Actes du colloque des 7, 8 et 9 juin 2012*, directed by Patrick Troude-Chastenet, 211-26. Paris: La Table Ronde, 2014.

6

Liberalism and the State in French and Canadian Technocritical Discourses

Intersections and Contrasts between George Grant and the Bordeaux School[1]

BY Christian Roy

> *Christian Roy is an independent scholar of intellectual and cultural history, an art critic and cinema critic, and a translator from several European languages, based in Montreal. A specialist of the French Personalist tradition, he published his thesis on its origins (PhD, McGill, 1993), having previously identified its Bordeaux "school" around Bernard Charbonneau and Jacques Ellul as an early fount of the critique of technology. He has just completed the first English translation of a book by Charbonneau (*The Green Light: A Self-Critique of the Ecological Movement, *Bloomsbury, 2018), soon to be followed by that of Ellul's posthumous* Theology and Technique *(Wipf and Stock). He is on the editorial committee of the* Ellul Forum *(ellul.org). In addition to articles on George Grant and many little-known Personalists, he has contributed entries on them to* The Encyclopedia of Modern Christian Politics *(Greenwood Press, 2006) and* Enciclopedia della persona nel Xx secolo *(Edizioni Scientifiche Italiane, 2008), in addition to writing* Traditional Festivals: A Multicultural Encyclopedia *(ABC-Clio, 2005). Many of his texts are available online at https://roychristian.academia.edu.*

* * * *

---

1. This article was originally a paper given at the Sorbonne in Paris on September 21, 2013, at the sixth Tensions of Europe Plenary Conference, "Democracy and Technology: Europe in Tension from the 19th to the 21st Century."

Technique as System, Technology as Ontology

IN ITS 1964 ENGLISH translation, Jacques Ellul's book on *The Technological Society* framed the definition of its topic in North America and beyond, even though its impact remained marginal in France, where it was first published in 1954. It was a belated fruit of over twenty years of critical reflection and activism in a Southwestern faction of the French Personalist movement, driven by Ellul's lesser-known mentor Bernard Charbonneau, who invented political ecology in that pre-war context.[2] Charbonneau (1910–1996) and Ellul (1912–1994) formed a tandem of thinkers who were so close that it almost did not matter which one of them discussed what topic; so much so that each devoted his first major book to the other's main concern. Having first originated the concept of technique as the distinctive, overarching organizing principle of modern society, Charbonneau entrusted it to Ellul, so that he, rather than this Christian anarchist, could dwell on the State in his own book *L'État*, which would only find a publisher forty years later, in 1987. It was around that time that the Société pour la Philosophie de la Technique was launched at the initiative of followers of Charbonneau and Ellul, the second generation of what may be seen as the Bordeaux School, by analogy with the Frankfurt School of critical theory.[3]

Not coincidentally, it was a pillar of the Société pour la Philosophie de la Technique: Daniel Cérézuelle, who, coming back from studying with Hans Jonas at New York's New School for Social Research, was the first scholar in France to discuss (alongside his own teacher) the Canadian philosopher George Parkin Grant (1918–1988), in a 1976 article for an early issue on technique in the journal *Les Études philosophiques* published by the Presses universitaires de France. Cérézuelle highlighted among the philosophical investigations of "the meaning and implications of technological progress" that had appeared in North America over the previous decade those that "tend to undermine the prevalent notion of the universality and axiological neutrality of the technological phenomenon," as the Bordeaux School had been doing since the early 1930s.[4] The parallel was left unmentioned in that text, but I want to explore it by following the thread of a line of argument Cérézuelle highlighted in Grant that can be traced back to Ellul, beyond the direct influence his book on *The Technological Society* had on the Canadian philosopher, as Grant himself readily admitted, e.g. in the

2. Roy, "Aux sources," 67–100.
3. Roy, "Ecological Personalism," 33–44.
4. English abstract of Cérézuelle, "La philosophie," 209.

effusive cover letter to the copy of his own *Technology and Empire* that he sent Ellul in gratitude when it came out in 1969.[5] It was indeed that very copy that Cérézuelle borrowed from Ellul (likely at the latter's suggestion) to write his review several years later.[6]

In his book, George Grant maintained that progressive narratives of emancipation were not really in a position to sustain a coherent challenge to the enfolding of all aspects of life within technology, which he defined as something more than technique as understood by Ellul as the whole complex of rational methods for absolute efficiency, since it entailed a "belief in the mastering knowledge of human and non-human beings." (In giving priority to a form of knowledge over a type of practice, Grant thus unwittingly came closer to Charbonneau's critical focus on Science over Technique as defined by Ellul.)[7] As both a practice and an ideology, Grant wrote, in passages quoted by Cérézuelle, technology "arose together with the very way we conceive our humanity as an Archimedean freedom outside nature, so that we can creatively will to shape the world to our values." The problem is then that "the moral discourse of 'values' and 'freedom' is not independent of the will to technology, but a language fashioned in the same forge together with the will to technology."[8] As a result, "our liberal horizons fade in the winter of nihilism" before "the pure will to technology (whether personal or public)"; for if, "within the practical liberalism of our past, techniques could be set within some context other than themselves—even if that context was shallow," "we now move towards the position where technological progress becomes itself the sole context within which all that is other to it must attempt to be present."[9]

---

5. I thank Jake Rollison for sending me a copy of this letter dated "6/5/1969," which he kindly brought to my attention after some research in the Fonds Jacques Ellul at the Institut d'Études Politiques in Bordeaux. See also Grant's "Review of *The Technological Society*," 413–18, and a Religious Studies dissertation highlighting Ellul's influence on Grant, e.g., on his joint critique of liberalism and technique: Muncaster, "Influence of Jacques Ellul."

6. As confirmed by Daniel Cérézuelle in an e-mail to the author on October 1, 2017.

7. See Charbonneau's original critique of Science as *Ultima Ratio*, paired with the reissue of his 1965 book, *The Paradox of Culture*, as its indispensable non-purposive alibi, in a diptych that may be seen as his putative answer to C.P. Snow's opposition of *The Two Cultures*: Charbonneau, *Nuit et jour*.

8. Grant, "In Defence of North America," in *Technology and Empire*, 32.

9. Grant, "In Defence of North America," in *Technology and Empire*, 40.

Ideological Correlates of Technology

Before Grant, the Bordeaux School viewed liberalism as the ideological seedbed of technology's threat to the values of freedom and equality claimed by that ideology. Ellul could describe "fascism as liberalism's child" in Personalist publications, for as Charbonneau had maintained early on in the newsletter of the Bordeaux group of Friends of the review *Esprit*, both, like communism, have quantifiable production as their final argument.[10] Fascism and communism, being but "spectacular reformisms," share in this the assumptions of the liberalism they aim to replace, and so cannot change an increasingly alienated daily life.[11] Grant also saw these three rival ideologies as the modern political systems consonant with the dominance of technology, which had replaced Christianity in Western man's assumptions about reality.[12] Asked about Ellul in a 1978 interview, Grant voiced his distaste "of the liberal and Marxist ideologists and their accounts of technology as a means at the disposal of human freedom. When they speak that way they forget that both capitalism and communism are but predicates of the subject, technology. Ellul's description of technology was quite outside such a shallow account, and he faced what was actually happening with his lucid French and Christian common sense.[13]

Ellul thus ascribed the emergence of a "pre-fascist mentality" to the fact that "*by proclaiming freedom of thought, liberal society had freed itself from thought*" since "*any thought is equivalent to any other*" and need not be matched by corresponding action to be validated. Subjective opinion and arbitrary imagination go unchecked, but remain powerless, while "*the material world tends to organize itself on bases that are absolutely independent of any effort of thought*";[14] until, that is, they are imposed as public dogma through advertising and propaganda, forming "abstract masses" of individuals whose psychological reactions are gauged and manipulated by the statistical methods of the social sciences. By its ability to go a step further and concretely mobilize these abstract masses, "fascism appears, from a social standpoint, as a better designed, more willful amorphism than the other, liberal state, but of the same nature, belonging to the same type of society[15]."

---

10. See Ellul, "Le fascisme," cited below and translated in this book as "Fascism, Son of Liberalism."
11. Charbonneau, "Les actes nécessaires."
12. Grant, "Religion and the State," in *Technology and Empire*, 41–60.
13. Schmidt, *George Grant in Process*, 146.
14. Ellul, "Le fascisme," 118–19.
15. Ellul, "Le fascisme," 136.

Even *"fascism's lack of theory is a liberal trait."*[16] Fascism is thus the worthy heir of liberalism: "it keeps all of its father's features—only with the addition of those of its mother, technique"—just as, for Grant, modernity itself, as "the dream of liberalism and its scientific mistress—'neutral' technology," seems destined to gut freedom and equality of substantive content.[17] Ellul concludes with the description of fascism he claims to find in Alexis de Tocqueville, when this nineteenth-century liberal thinker, who remained a touchstone for both Charbonneau and him, writes of "democratic societies that are not free though they may be rich, refined, ornate, magnificent even, powerful by the weight of their homogeneous mass," where private virtues may still flourish even in the absence of civic spirit, once this mass quietly embraces absolute rule.[18]

The Liberal Banalization of Evil

In a 1968 collection of "candid Canadian opinions" of the United States, Grant used that same example to likewise "assert the ancient and forgotten doctrine that evil is, not the opposite, but the absence of good."[19] This was fostered by liberalism's "value-freedom" as theorized by John Rawls, of whose *Theory of Justice* he was thus an early critic, long before the communitarians.[20] "The emptiness of a moral tradition that puts its trust in affluence and technology results in using any means necessary to force others to conform to its banal will," "when deemed necessary to comfortable self-preservation," in a "use of power" "which perpetrates evil from its very banality."[21] For "the 'good life' to which it is proper to aspire in technological society is not a life constrained by moral judgments. . . . This quest for freedom divorced from virtue entails the desire to dominate necessity, hence leads to tyranny."[22]

Charbonneau already saw the banality of evil as an issue going far beyond the specific "Responsibilities of the German People" he discussed in a November 1945 article for one of the Protestant publications his

16. Ellul, "Le fascisme," 118.
17. Ellul, "Le fascisme," 136; Flinn, "George Parkin Grant," 199.
18. Ellul, "Le fascisme," 137.
19. Grant, "From Roosevelt to LBJ," 468.
20. Grant, *English-Speaking Justice* (first published by Mount Allison University, Sackville, New Brunswick, as the Josiah Woods Lectures, 1974).
21. Grant, "From Roosevelt to LBJ," 468.
22. Badertscher, "George P. Grant," 84, summarizing Grant's essay "Tyranny and Wisdom" (*Social Research* 1964) in *Technology and Empire*, 79–109.

friend Ellul gave him access to, agnostic though he was. "If we can only imagine a mechanical civilization where personal responsibility is lost," then "we will have to manufacture good Germans the same way Hitler manufactured bad Germans. But let us remember that it is when we start from those neutral techniques that can be used for anything indiscriminately, when we start especially from this neutral being that gets formed and deformed, that everything is possible,"even when it is a liberal regime that proposes to "win hearts and minds"—or else.[23] Thus, in 1967, Grant is not surprised that "what is being done in Vietnam is being done by the English-speaking empire and in the name of liberal democracy," and not by what "could be seen as the perverse products of western ideology—National Socialism or communism."[24]

Charbonneau presciently picked up on a tell-tale early sign of that shift within liberalism in a 1952 article on this "Heart-Rending Revision" for the Protestant weekly *Réforme*. He argued that Western societies, "particularly Anglo-Saxon ones, were founded on the myth of Progress that confused material progress and spiritual progress, that of collective power: of science and technology, with that of individual freedoms. There wasn't a problem: it is understood that that the societies that are technically most advanced are also the freest, as shown by the case of America." "Having long confused Progress with Freedom and Democracy, America is now mulling over their contradiction, but I fear it won't be for long," for "today, it is becoming perfectly natural to sacrifice the latter to the former, since the facts have demonstrated that Freedom is an obstacle to Progress," in the guise of "totalitarian successes." Identifying their values with their national power, when forced to choose, "liberal democracies will brutally suppress their political freedoms, equality in education or salaries, leading to a regime where the dictatorship of the central power would underwrite a policy of massive investments," surviving freedoms having first been emptied of content by the cult of efficiency. "While Human Rights are on display on the first floor, torture is being practiced in the basement"—be it in Algeria at that time or in Guantanamo in ours. For whether it be H-bombs or drones, "what is the use of changing your weapons system without also updating your principles," as Charbonneau had first asked upon introducing the musings of "an American journalist" on which this text was a commentary, to the effect that "we have to wake up from our illusions of easy technical and material superiority." Soviet life is based on force rather than consent, but "are we so

---

23. Charbonneau, "Responsabilités du peuple allemand," 85–86.
24. Grant, "Canadian Fate and Imperialism," in *Technology and Empire*, 65.

sure that our social aims, derived from the individual's right to free will, are stable, constructive and based on lasting values?"[25]

Canadian Ambivalence Toward America as Technology's Empire

The author of this quote, identified as Lester Pearson, was actually neither American, nor a journalist, but Charbonneau still could not have chosen a better specimen of the contradiction at the core of Anglo-Saxon liberalism than this Canadian minister of External Affairs who would win the Nobel Peace Prize in 1957 for his invention of UN peace-keeping troops during the Suez Crisis, and would go on to become leader of the Liberal Party in 1958 and Prime Minister from 1963 to 1968. The policy of military, even nuclear cooperation with the United States that brought Pearson to power was the pretext for the book that made Grant famous in his own country in 1965, *Lament for a Nation: The Defeat of Canadian Nationalism*. As Grant explained in introducing its 1970 reissue, behind the specific political decisions arising from Canada's ambiguous status within the American empire was "the deeper question of the fate of any particularity in the technological age. What happens to nationalist strivings when the societies in question are given over, at the very level of faith, to the realisation of the technological dream? At the core of that faith is service to the process of universalization and homogenization" in the name of technology's "one best means."[26]

Hence a Canadian sensitivity to this issue, exemplified by Grant among others,[27] since any "distinction will surely be minimal between two nations which share a continent and a language especially when the smaller of the two has welcomed with open arms the chief instrument of its stronger brother—the corporations."[28] Viewing the United States as "the only society which has no history (truly its own) prior to the age of progress,"[29] and as a result, no horizon beyond the one defined by technology, Grant lamented the passing of a British North America that drew from its acknowledged roots in the

25. Charbonneau, "Révision déchirante," in *Réforme*, from a clipping dated December 1952 without further identifying data, that was shown to me by the author's widow Henriette around the turn of the century. Charbonneau would go on to publish numerous essays in this Protestant periodical (which still exists) over the following decade.

26. Grant, *Lament*, lxxii.

27. See Kroker, *Technology and the Canadian Mind*.

28. Grant, *Lament*, lxxii.

29. Grant, "In Defence of North America," in *Technology and Empire*, 17. On this much-debated "Red Tory" paradigm of Canadian identity, often associated with George Grant, see Horowitz, "Conservatism."

older European cultures of France and the United Kingdom the "belief that on the northern half of this continent we could build a community which had a stronger sense of the common good and of public order than was possible under the individualism of the American capitalist dream" unleashed by the Revolution his Loyalist ancestors had fled.[30]

Grant sympathized with French Canadian nationalism for keeping a similar hope alive, despite its current modernizing wager to have it both ways. He saw this as a typically Canadian position and thought it had "been put most absurdly by the Liberal leader in Quebec, M. [Robert] Bourassa: 'American technology, French culture'—as if technology were something external (e.g. machines) and not itself a spirit which excludes all that is alien to itself. As Heidegger has said, technique is the metaphysic of the age."[31] Feeling that a strong national State was the only thing that might defend Canada's identity and communitarian ethos against the encroachments of American corporate liberalism, Grant admired Charles De Gaulle for taking such a stance for France, and giving his country a measure of independence from the dictates of the United States as the hegemonic center of the liberal version of the "universal homogeneous State" devoted to neutralizing "politically relevant natural differences among men" "by progressing scientific technology," "thanks to the conquest of nature and to the completely unabashed substitution of suspicion and terror for law," in the terms drawn from Leo Strauss's debate with Alexandre Kojève that Grant applied to America.[32]

Critiques of Technology from the Margins of Modernity

Charbonneau, on the other hand, could never forgive General De Gaulle for making France into a nuclear power, and presiding over the planned modernization of the country, justified by the bid to retain some status on the world stage. For in the name of "a certain idea of France" (to use De Gaulle's famous phrase), the reality of the country, and whatever was worth preserving about it, was being readily sacrificed, from the age-old nature-culture synthesis of the countryside down to the nation's very existence and that of all mankind as a likely result of nuclear proliferation and the increasing risk of worldwide conflict. This for him exemplified the logic of the modern State as it has developed in the West since the eleventh century as

30. Grant, *Lament*, lxxiii.
31. Grant, *Lament*, lxxii.
32. Strauss, *What Is Political Philosophy*, 132–33, cited in Grant, "Tyranny and Wisdom," in *Technology and Empire*, 96.

the centralizing vortex of the converging control processes culminating in technology.[33] Ellul also underlined that "the increasing interrelationship of state and technique affects political life on a global level. The ultimate product is a total world civilization."[34] Grant would have agreed that "protecting romantic hopes of Canadian nationalism is a secondary responsibility" "in an age when the alternatives often seem to be between planetary destruction and planetary tyranny,"[35] feeding the dialectic of system and chaos that Charbonneau, in a book written between 1951 and 1967, described as the driving force of exponential development, in a vicious cycle calling on ever more technological control to counter the latter's increasingly disruptive environmental and social effects.[36]

For Charbonneau and Ellul, any nation-state, including such smaller-scale ones as might result from the breakup of larger units, was bound to be a vector in that worldwide process of technological homogenization, whatever claims of cultural particularity might be invoked to justify building a State apparatus so as to be politically and economically competitive. That is why, shunning the draw of Paris and faithful to their provincial roots, they took aim at the hold of the centralized State in France as the oldest modern nation, in a defense of local life against planned modernization and untrammeled development that happened to be rooted in the same Southwestern region as the Girondin party of federalists crushed by the Jacobins in the French Revolution.[37] Faced with a French centralism whose claim to embody the common good went unchallenged, Charbonneau appreciated what remained of individualism in Anglo-Saxon cultures, as it was this Protestant element that had allowed them to discover nature as an ally for individuals who resisted the encroachments of industrial society and the technocratic State.[38] Conversely, Grant liked to turn to France for a sense of the common good such as he was hoping to maintain through Canadian statehood, in the face of American corporate domination built on liberal assumptions about the innocence of technology and the possessive individualism it enabled. Yet it seems no coincidence that the powerful critiques of technique's alleged neutrality mounted first by the Bordeaux School and later by George Grant arose on the marches of France and the United States respectively as the historic centers of liberal progressivism in the Old and New Worlds, motivated

33. Roy, "Charbonneau et Ellul," 291.
34. Ellul, *Technological Society*, 318.
35. Grant, *Lament*, ix–x.
36. Charbonneau, *Le Système et le Chaos*.
37. See Charbonneau, *Sauver nos régions*.
38. See Charbonneau, *Green Light*, chapter 6, "Nature and Christianity."

by concern for the fate of both local particularity and genuine personal freedom in the Brave New World remade as one by technology. For they all saw in technique the underlying dynamics shared by the liberal consensus with overtly State-worshipping ideological competitors, even if the former came to prevail as both its matrix and its most potent vector.

Bibliography

Badertscher, John. "George P. Grant and Jacques Ellul on Freedom in Technological Society." In *George Grant in Process: Essays and Conversations*, edited by Larry Schmidt, 79–89. Toronto: House of Anansi, 1978.

Cérézuelle, Daniel. "La philosophie de la technique en Amérique." *Les Études philosophiques* 2 (1976) 209–22.

Charbonneau, Bernard. "Les actes nécessaires." *Bulletin du groupe de Bordeaux des Amis d'Esprit* 2 (s.d.).

———. *L'État*. Classiques des sciences sociales series. Paris: Economica, 1987.

———. *Le Feu vert. Autocritique du mouvement écologique*. Intr. D. Cérézuelle. Lyon: Parangon/Vs, "L'Après-développement" series, 2009. (Original edition: Paris: Karthala, 1980) English translation: *The Green Light. A Self-Critique of the Ecological Movement*. Translated by Christian Roy. London: Bloomsbury, 2018.

———. *Nuit et jour. Science et Culture*. Classiques des sciences sociales series. Paris: Economica, 1991.

———. "Responsabilités du peuple allemand." *Le Semeur* 2/1 (Nov. 1945) 81–86.

———. "Révision déchirante." *Réforme* (December 1952).

———. *Sauver nos régions. Écologie, régionalisme et sociétés locales*. Les Dossiers de l'écologie series. Paris: Le Sang de la Terre, 1991.

———. *Le Système et le Chaos: où va notre société?* La pensée écologique series. Paris: Le Sang de la Terre, 2012. (Originally published as *Le Système et le Chaos: Critique du développement exponentiel*. Paris: Anthropos, 1973).

Ellul, Jacques. "Le fascisme, fils du libéralisme." *Cahiers Jacques Ellul. Pour une critique de la société technicienne* 1 (2003) 113–37. (Originally published in *Bulletin du groupe de Bordeaux des Amis d*'Esprit 4 (s.d.), then in *Esprit* 53 (1937) 761–97.)

———. *The Technological Society*. Translated by John Wilkinson. New York: Vintage Books, 1964.

Flinn, Frank K. "George Parkin Grant: A Bibliographical Introduction." In *George Grant in Process: Essays and Conversations*, edited by Larry Schmidt, 195–99. Toronto: House of Anansi, 1978.

Grant, George. *English-Speaking Justice*. Toronto: House of Anansi, 1998.

———. "From Roosevelt to LBJ." In *Collected Works of George Grant*. Vol. 3, *1960–1969*, edited by Arthur Davis and Henry Roper, 466–69. Toronto: University of Toronto Press, 2005. (Originally published in *The New Romans: Candid Canadian Opinions of the U.S.*, edited by Al Purdy. Edmonton, AL: M. G. Hurtig, 1968.)

———. *Lament for a Nation: The Defeat of Canadian Nationalism*. 40th Anniversary ed. Montreal: McGill-Queen's University Press, 2005.

———. "Review of *The Technological Society*, by Jacques Ellul." In *Collected Works of George Grant*. Vol. 3, *1960–1969*, edited by Arthur Davis and Henry Roper, 413–

18. Toronto: University of Toronto Press, 2005. (Originally published in *Canadian Dimension* 3/3–4 (1966) 59–60.)

———. *Technology and Empire: Perspectives on North America*. Toronto: House of Anansi, 1969.

Horowitz, Gad. "Conservatism, Liberalism, and Socialism in Canada: An Interpretation." *Canadian Journal of Economics and Political Science* 32/2 (1966) 143–71.

Kroker, Arthur. *Technology and the Canadian Mind: Innis/McLuhan/Grant*. Montreal: New World Perspectives, 1985.

Muncaster, Andrew. "The Influence of Jacques Ellul, Martin Heidegger and Simone Weil on George Grant's Changing Understanding of Technology." PhD diss., McMaster University, 2008.

Roy, Christian. "Aux sources de l'écologie politique: Le personnalisme gascon de Bernard Charbonneau et Jacques Ellul." *Canadian Journal of History* 27/1 (1992) 67¬100.

———. "Charbonneau et Ellul, dissidents du 'Progrès'. Critiquer la technique face à un milieu chrétien gagné à la modernité." In *Une autre histoire des "Trente Glorieuses." Modernisation, contestations et pollutions dans la France d'après-guerre*, edited by Christophe Bonneuil, Céline Pessis, and Sezin Topçu. Paris: La Découverte, 2013.

———. "Ecological Personalism: The Bordeaux School of Bernard Charbonneau and Jacques Ellul." *Ethical Perspectives* 6/1 (April 1999) 33–44.

Schmidt, Larry, ed. *George Grant in Process: Essays and Conversations*. Toronto: House of Anansi, 1978.

Strauss, Leo. *What Is Political Philosophy*. Glencoe, IL: Free Press, 1959.

7

Jacques Ellul and Charles Taylor on the Sacrality of Secularism

BY Gregory Wagenfuhr

> *Gregory Wagenfuhr is author of* Plundering Egypt: A Subversive Christian Ethic of Economy *(Cascade, 2016). He currently serves as a theology consultant to the Covenant Order of Evangelical Presbyterians (ECO). He earned his PhD in Theology and Religious Studies from the University of Bristol (UK) researching Jacques Ellul, the sacred, and Christian faith as desacralization. His current projects include* Unfortunate Words of the Bible and Christianity, The Creation of a Theory of Value, *and* The Cross and the Scales: A Subversive Christian Theology of Justice.

* * * *

MUST SECULARITY BE SACRALIZED for the establishment and maintenance of a secular society? This chapter contrasts two visions of the secular age, its problems, and its relation to the sacred. The first scholar we consider is Jacques Ellul (1912–1994), who in the context of this volume needs no further introduction. We will be concentrating primarily on his *New Demons*, though much of his other work plays into the arguments of this chapter. The second scholar is Charles Taylor (b. 1931), a highly celebrated contemporary philosopher. His prize winning *A Secular Age* explores the historical development of modern secular perspectives.

Defining the Sacred

The sacred can be defined in a number of ways and this is not the place to provide all the options. For other options see the bibliography. The sacred is an essential ingredient in the social construction of reality, a process called *cosmisation*.[1] This is not an ontological statement about the universe as such, but about how social groups create a world of meaning, both being formed by their environment, and forming their environment. I think we can take Charles Taylor's notion of "social imaginary" as a rough equivalent. Taylor is not deeply concerned with the sacred, noting that it is more of a French scholarly device.

For Ellul, the sacred is the most foundational and unquestionable value that creates a topography for social reality and meaning. He calls this "absolute value" in an unfortunate use that does not align with that term's usage in the philosophy of value, axiology. For Ellul, the sacred comes about through a process of integrating oneself into an imposed reality by imagining that order to be the condition of life.[2] Sacralization, then, is the unconscious attribution of ultimate and highest value. I will follow him in this definition.

Defining Secularity

Secularity is a term with many usages, and it is usually described by way of what Taylor calls a "subtraction story," in which modern humans threw off the shackles of superstitious religious belief in the divine, which liberated us to understand the world in a purely profane, or immanent way.

As both Ellul and Taylor show in their respective books, that narrative is historically false, and conceptually inaccurate. Though, as Taylor notes, false stories are often more historically influential than true ones for the role they play in forming beliefs. So, in order to understand secularity, we follow Taylor in his division of "secularity" into three definitions:

- Secularity 1: Under this definition, public spaces must be emptied of reference to God or transcendent reality. This is what is often meant by "secular" in political or scientific discussions. This is legislated and dogmatic secularity.
- Secularity 2: This refers to the decline of religious belief, practice, and church membership. This notion of secularity is not about legislation, but about the transformation in the conditions of religious adherence.
- Secularity 3: Here belief in God is one option among others, and frequently not the easiest to embrace. This refers to the social imaginary, to the framework and background that gives rise to secularity 1

---

1. Cosmisation is a term used by Peter Berger in his *Sacred Canopy* as he discusses the social construction of space and time.
2. Ellul, *New Demons*, 57.

as a rational expression and results in secularity 2. This is the type of secularism that interests Charles Taylor, as it speaks to the felt reality of a people.[3]

Most discussions of secularism refer to secularity 1. The common subtraction story is an ideology of secularity 1 without reference to the development of its conditions of plausibility. Taylor's aim in *A Secular Age* is to describe how secularity 3 came about, such that secularity 1 is not only plausible, but assumed. Taylor's guiding question is, "Why was it virtually impossible not to believe in God in, say, 1500 in our Western society, while in 2000 many of us find this not only easy, but even inescapable?"[4]

Jacques Ellul wrote extensively on the modern age, though not often from the perspective of secularity. Ellul in many ways prefigured some of Taylor's criticisms of "subtraction stories" as well as the confusion about defining exactly what secularity means. For Ellul, most arguments about the joys of secularism are based in a convenient confusion that enables a circularity of logic. This confusion is between a "moral doctrine" and a factual situation. By alternating between these two, authors are able to claim that the doctrine is being confirmed by reality, but this is a reality created by the moral doctrine in the first place.[5] Translating this into Taylor's terms, Ellul is identifying secularity 1 and secularity 2, and like Taylor, offers narratives that explain secularity 3.

Ellul would not, however, define the modern social imaginary primarily by its secularity. Indeed, he is rather dismissive of the notion of secularity, as though there were a possible society without sacral structures, beliefs, and myths. The twentieth century was a time of high sacralization, Ellul believed, of the nation-state, and of technique. These two powers represent the defining features of the modern human situation, such that Taylor's secularity 3 is better understood as the necessary result of the elevation of *la technique* and the nation-state to sacral status. In that sense, secularity 3 is really a subset of a social imaginary defined by technique, a subject Taylor largely bypasses.

Already, then, we can see that the projects of Ellul and Taylor run parallel in many ways, both seeking to understand the deeper currents of modern doctrines, though each coming to rather different conclusions and offering different narratives. As we will see, these differing narratives end in very different places.

3. Taylor, *Secular Age*, 2–3.
4. Taylor, *Secular Age*, 25.
5. Ellul, *New Demons*, 32.

Ellul's Modern Sacred: The Necessity of Sacrality

In *New Demons*, Ellul offers a narrative of the contemporary world, beginning with early Christendom. Christianity had a unique development in the history of religions, as it did not develop organically as an embedded feature of a social group. This means that the adoption of Christianity by the Roman state was a conscious choice, as was the further transformation of a social system. Indeed, Christendom invented the concept of a social order consciously malleable by human choice and consent. This means that it advanced rationally and with great resistance. Ellul is critical of overly simple readings of history in which Christianity simply baptized paganism.[6] The historical evidence suggests that people rationally attempted to apply their newly chosen faith to all aspects of life. Their religious backgrounds were Christianized, as they (like all people) lacked the ability to make a radical imaginative leap and transform everything instantly by their new faith. Thus, Christendom was less about the church's union with the state. Instead, Christendom was defined by the attempts Christians made to rationally transform their social imaginaries with their new, all-embracing faith.

But this had serious consequences. Christianity had to become an ideology to be all-embracing. And thus, not everyone adopted it as a personal faith. Christendom, then, was more about external affiliation than internal belief. This had the result that the gospel was "translated to a lower register."[7] Also as a consequence, morality, which is external, became the defining feature of adherence, and thus also the personal devotion of the faithful. But perhaps most vital, Ellul thinks, in the rise of Christendom was that the decline of Rome left a vacuum of basic human services that the church stepped into.[8] The church, as universal, also became a more fundamental basis of unity than state allegiance. As all these things happened, we might say, Christianity became re-embedded and subverted. But this was not the same as it had been for paganism. It was not an organic embeddedness, but one that had been adopted.

When it comes to post-Christendom, Ellul is critical of those who follow Bonhoeffer's famous concept of "man come of age." This is ignorant of history, Ellul thinks, and is doctrinaire. It is the result of transforming a wide cultural acceptance of an ideology into a value judgment on a human reality. Ellul does not give a long social history of the development of secularism, as Taylor does. He is more concerned with refuting the false

---

6. Ellul, *New Demons*, 4–6.
7. Ellul, *New Demons*, 7.
8. Ellul, *New Demons*, 9–10.

justifications of those who claim that this is a more advanced, adult, stage of civilization. He demonstrates this by examining how the social functions of more traditional religions have migrated to other areas, such that the twentieth century up to the 1970s represented a deeply sacralized time. These functions fulfilled by the sacred are broadly understood as orientation and integration into a milieu.[9] Ellul will elsewhere say that *la technique* is the modern environment.[10] The state and *la technique* have created a reality far different from any that had gone before, thanks to global systems and technological developments.

The sacred today, Ellul believes, works a bit like a map. Its two axes are defined by four cardinal, sacred points. These are the nation-state/revolution, and *la technique*/sex. The nation-state is opposed by sacralized revolution, and *la technique* is opposed by sex. We don't have the space to argue for or against the merits of this identification. But what is important to grasp is that Ellul believes that the sacred has migrated from religion into the nation-state and into *la technique*. Thus, modern society is not at all desacralized or freed from the needs of the justification that religion once provided. All the evidence he needs is conveniently supplied by daily life and news in the mid-twentieth century, and much of his other work on *la technique* goes a long way toward demonstrating the sacral nature of technology. The ideologies and realities of Maoism, Stalinism, and Hitlerism show how atheistic regimes adopt sacral attitudes and religious forms. But Ellul also sees evidence of this is modern democracies as well, and in consumer enthusiasm for technological gadgets.

Importantly, for Ellul, people are responsible for this sacralization. But it could also not be otherwise. We live in the most religious of all times, Ellul thinks. But it's not because humans are by nature religious creatures, rather, the modern situation demands it. He believes a non-religious, adult situation is possible in theory. But "Man is absolutely not free to sacralize or not to sacralize technology. He cannot keep from reconstructing a meaning for life on that as a basis."[11] The realities of *la technique* demand sacralization, and thus, neither technology, nor the state are the means by which humans will attain an adult life living in a reality that is not socially constructed and self-legitimating.

For Ellul, the answer lies in desacralization. For the nation-state he does not advance revolution, which only confirms the necessity and essential reality of the sacralized state, but a kind of peaceful anarchy, the negation of the

---

9. See my PhD thesis, "Revelation and the Sacred Reconsidered," for far more detail.
10. Ellul, *Technological System*, 34–50.
11. Ellul, *New Demons*, 206n2.

state. And for *la technique*, Ellul does not offer a radical celebration of the body, which only confirms the sacral dualistic value of disembodiment in technology. For Ellul, only the revelation of God in Jesus Christ can desacralize this sacred of *la technique*, and lead to an adult situation.

Taylor's Secular Age: Avoiding Closed World Systems

For Taylor, secularity 3 arose by a number of historical factors. The first factor, and one that continues throughout the narrative in various guises, is reform. Reform movements, with the most obvious example being the Protestant Reformation, attempt to rationally transform society. Reform in the late medieval period was characterized by disenchantment of reality and "church magic." Reform disembeds religion, because it critiques a present reality from a rational point of view, whether theological or not. The Reformation was a theological movement of reform that used faith to critique particular aspects of embedded religious reality. This had important results, an inverse image of what Ellul described in the rise of Christendom. Disenchantment meant that religious commitment had to become internal, and thus more fervent. This called into question what Taylor calls a "two-speed" system in which the professional clergy and the laity had entirely different expectations of religious commitment. This was legitimized by the social complementarity that characterized medieval and more traditional societies, in which people in different vocations supported one another. The Reformation eliminated the clergy/laity divide, thus undermining complementarity, which meant that Protestantism called for the average person to attain much higher level of personal devotion, like that of the clergy.

Taylor also observed in reform the increasing systematization of the world. Taylor talks of the "disciplinary society" in which society should be subjected to rational scrutiny. This produced critiques of the warlike aristocracy, and began replacing a militarized society with an economic society grounded in theories of a rational systematic economy, as in Adam Smith. This, of course, came after a time in which the universe itself was subject to rational systematization, as with Newton. What began with the elite eventually permeated through all ranks, and eventually this disenchanting "rage for order" and discipline came to all classes.

Taylor sees disenchantment as a factor of the Axial revolution (ca. 800–200 BCE), which Christianity brought to its completion only in the 1700s. In opposition to this embedded way of life characterized by orders of complementarity, modernity is characterized by what Taylor calls the "Modern Moral Order." Though more complex than a single sentence, broadly this means that society believes in disciplined order guided by rules, working for personal thriving, with the result of mutual benefit. The individual comes first, and if the individual is taken care of, society as a whole will benefit. This is the inverse of embedded society in which society comes first.

The nineteenth century was characterized by the "nova effect" in which belief, as a personal option, gave rise to an increasing diversity of beliefs and self-expressions. This was confined to the cultural elite. World War I translated it to the masses. This gave rise to what Taylor identifies as the "supernova" in which these options have expanded beyond counting. Thus, we live in the "age of authenticity" in which self-development and expression are most important. Religion is a personal choice, not a factor of embeddedness.

But all of this is driven by a common narrative that forms the social imaginary. Although options for belief and moral expression are now highly varied, they must be motivated and held together in some way. For Taylor this modern social imaginary plays upon the subtraction stories of secularity 1. Authenticity is believed to be achieved by the abandonment of religious superstition, of transcendence. There is nothing but the immanent, and to believe otherwise becomes dangerous to the modern moral order, or so the common narrative goes. But this is a false dogmatism, Taylor thinks. The immanent frame can have two "spins": open and closed. It allows for belief in transcendence or for disbelief. But it demands neither, and should not dogmatically require others to adhere to either.

Taylor engages in a deconstructive reading of "Closed World Systems" (CWS) and their belief that a closed reading is natural. He demonstrates that CWS is driven by value and ethics rather than by epistemology. Science doesn't show that God is dead, as it were, but the atheistic value uses the epistemological excuse to justify belief. The same is just as true of belief in transcendence, though this has been well rehearsed ever since Feuerbach.

In spite of all attempts to the contrary, we all have master narratives, Taylor avers. They are essential to human thinking. But we need to be clear about what we're doing, and not pretend that we are giving a neutral or natural reading of history.[12] Rather, value is playing a decisive role in the subtraction story and using science (which is an epistemological method that creates/"discovers" ontology) to legitimize these values in a way that does not rationally follow. "Coming of age" and the "death of God" are subtraction stories that have become Wittgensteinian background pictures for our society.

12. Taylor, *Secular Age*, 573.

Ellul and Taylor are offering similar portraits of the contemporary social imaginary. Both agree that our Western society is suffused with false beliefs about its own maturity brought about by secularity 1. Both show that the narratives offered by secularity 1 are unconvincing and myopic. And both suggest that it is theoretically possible to live in an immanent space open to transcendence. Neither desire to return to some imagined ideal state.

Ellul offered anarchy and Christianity as desacralizing powers. For Taylor, scapegoating violence is self-evidently problematic, and is a "virus" present in both the transcendent perspective and the immanent perspective.[13] He offers three key questions of a program for new common ground: finding moral sources to justify our belief in universal human rights; finding how to defeat scapegoating violence that occurs equally in immanent-only and transcendent-only perspectives; and finding how to reaffirm what has been lost by the mutilations that movements of reform have necessarily imposed upon us, and that modern spiritual expressions and conversions seek to recapture.

Must the Secular Be Sacred?

Subtraction stories are similar to justificatory myths that sacralize the secular. Value is the key to linking the insights of Ellul and Taylor. For Ellul, the sacred is the highest of values. The sacred represents foundational and unquestionable values, such that myths must be created to justify them, for they cannot stand rational scrutiny.[14] And for Taylor, values drive the supposedly epistemological subtraction stories about "coming of age" and the "death of God." Taylor's account of the modern moral order of mutual benefit is, for him, something that cannot really be changed, but that does not require a doctrinaire belief in immanence-only, or in transcendence. Instead, it requires that we find new common ground, and new ways to legitimately experience the full range of human flourishing without engaging in the mutilations and scapegoating violence all those who advance a necessary narrative.

Thus, for Taylor the secular does not have to live within subtraction stories. It is possible to conceive of an immanent world open to transcendence. Something of the past can be recovered without anachronism or adopting ancient perspectives. For Ellul on the other hand, the secular must be sacralized because of the nature of la technique, as it has created an entire environment that characterizes the very expressions of human thought.

13. Taylor, *Secular Age*, 709.
14. Ellul, *New Demons*, 55.

Taylor seems to be saying that there is hope in a future open secular society, and that real progress is possible, provided we reform our movements of reform, as it were. Ellul can see only continual and deeper alienation by *la technique* and the nation-state.

The calm tone of Taylor and his many accolades stands in opposition to Ellul's more confrontational writing style and smaller enduring audience. But this conveniently hides some vital problems with Taylor's treatment of Western history. He says little to nothing about technology's involvement with the story, or with the immense ecological destruction it has brought about. This is a major lacuna in Taylor's narrative, because it suggests that our hopes of solving human social problems by being more open, humble, and responsible falls flat in the face of the immense problems facing the twenty-first century. Ellul's broad corpus and his critique of the sacralization of *la technique* provides an important voice of dissent against false solutions by misidentified problems.

That's not to say that Taylor's narrative is wrong. However, the common ground he seeks is not possible without sacralization and its attendant problems. All forms of unity require reference to common value, which must have some genesis, some background, some common and sacralized environmental condition. Granted, if we could be mature enough to recognize our moral motives, as Taylor suggests, we wouldn't have to reify this transcendent value. But evidence is lacking for any widespread human ability to be this transparent about moral motives. Transcendent value, historically, is sacralized. Given that both Taylor and Ellul demonstrate how even those who attempt rational and profane valuation end up engaging in sacred ordering and violence (like Robespierre), we would be wise to be skeptical about future possibilities to the contrary. It would follow that either this is a necessary condition of human unity qua human, or as an accidental, though ubiquitous occurrence.

I contend that Ellul's answer is the correct one and the corporation and artificial intelligence provide evidence that humans have little in the way of maturity to show for their technical progress. The corporation is an excellent example of modern value reification. The corporation's nature as a *persona ficta* enables the abandonment of personal responsibility by projecting it onto an imagined being. This, of course, enables greater strategic risks to be taken, and thus the creation of more wealth more efficiently. The corporation, I have argued, is the first conscious reification in human history. As such, and given that corporations are now so ubiquitous that nearly every object bears their logo, we live in an entire milieu or cosmos in which reality is more and more defined by fictive beings who exist to prevent personal responsibility. I can think of no better Feuerbachian or Nietzschean definition of religion and its problems.[15] I would also suggest that virtual reality has the same ontological reality as the corporation and is, I would argue, its natural growth and extension to encompass an entire created cosmos in which one may spend the bulk of one's life.

Artificial intelligence offers another important perspective on this discussion. For, as Taylor noted, the common CWS narrative purports to be about epistemology and ontology, but these are really a cover for values. The development of AI demonstrates the desire for the creation of a perfect, transcendent other, and for the total instantiation of a set of values masked in an artificial being developed by artificial epistemology ("machine learning"). If such a reading of the contemporary situation is valid, it ought to be a frightening matter, since sacral violence is so prevalent as a result of firm belief in a value system masked by myths of rationality.

In short, Taylor's optimism seems dangerously ignorant of the escalation of weapons available for future scapegoating violence that will likely bear no reference to traditional gods, but all too much reference to the immense proliferation of transcendent beings in our time. The reality is, everyone believes in transcendence, because this is the only basis for shared value. The real question is the ontological reality of the transcendent other. Secularity 1 is really a doctrine against the ontological and personal reality of the transcendent, but it in no way abandons transcendent value, nor ontologically real but impersonal transcendent beings.

Conclusions

For Ellul, Christian faith as revealed in Jesus Christ offers the only lasting possibility for continual desacralization. Ellul's narrative of post-Christendom allows for Christian faith to return to its pre-subverted forms and thus allows its deeply critical voice to speak more clearly. In a sense, Christian faith is well placed in the twenty-first century to offer a valid alternative. It offers a legitimately external voice through revelation, Ellul believes.

For Taylor, the dream of an open secular world offers freedom of expression without recourse to violence and mutilation. Human flourishing is something we all seem to value in the modern West, so if we can find common moral ground for it, we can pursue it well. In my view, that common ground already exists through the narratives that bring about our modern understanding of human flourishing as material wealth, i.e. technology and the corporation. And yet, these are the very irresponsible sacral orders that

15. See Wagenfuhr, *Plundering Egypt*.

are responsible for the imminent destruction of the human species within a century, if Stephen Hawking is to be believed.

We may at least conclude with this, Jacques Ellul offers a unique voice and critique that must be heard again today. Challenging the narratives of masked sacral values is a vital activity. This is a movement of reform, and we must ask the world to consider whether human flourishing as we currently understand it has made the earth itself a scapegoat, and whether "mutilating" ourselves by abandoning our current social imaginary is wrong if it results in the survival of the species and thriving of the natural world.

Bibliography

Berger, Peter L. *The Sacred Canopy: Elements of a Sociological Theory of Religion*. New York: Anchor, 1990.

Caillois, Roger. *Man and the Sacred*. Translated by Meyer Barash. Urbana: University of Illinois Press, 2001.

Eliade, Mircea. *A History of Religious Ideas: From the Stone Age to the Eleusinian Mysteries*. Translated by Willard R. Trask. Chicago: University of Chicago Press, 1978.

Ellul, Jacques. *The New Demons*. Translated by C. Edward Hopkin. New York: Seabury, 1975.

———. *The Subversion of Christianity*. Translated by Geoffrey W. Bromiley. Grand Rapids: Eerdmans, 1986.

———. *The Technological Bluff*. Translated by Geoffrey W. Bromiley. Grand Rapids: Eerdmans, 1990.

———. *The Technological Society*. Translated by John Wilkinson. New York: Knopf, 1964.

———. *The Technological System*. Translated by Joachim Neugroschel. New York: Continuum, 1980.

Girard, René. *Violence and the Sacred*. Baltimore: Johns Hopkins University Press, 1977.

Lynch, Gordon. *The Sacred in the Modern World: A Cultural Sociological Approach*. New York: Oxford University Press, 2012.

Taylor, Charles. *A Secular Age*. Cambridge: Belknap Press of Harvard University Press, 2007.

———. *Modern Social Imaginaries*. Durham: Duke University Press, 2004.

Wagenfuhr, G. P. *Plundering Egypt: A Subversive Christian Ethic of Economy*. Eugene, OR: Cascade, 2016.

———. "Revelation and the Sacred Reconsidered: The Revelation of God in Jesus Christ as Desacralising Reorientation to 'Milieu' in and beyond Jacques Ellul." PhD diss., University of Bristol, 2013.

8

Illusion, Revolt, and Revolution

BY Samir Younés

> *Samir Younés is Professor of traditional urbanism, architecture, and architectural theory at The University of Notre Dame where he was Director of Rome Studies and Director of Graduate Studies in architecture. He has lectured around the globe at the Universities of Bologna, L'Ecole des Beaux-Arts, Paris, the University of Portsmouth (UK), and elsewhere. Among his many publications are* The Imperfect City: On Architectural Judgment *(Ashgate, 2012; Routledge, 2016) and* The True, the Fictive and the Real, Quatremère de Quincy's Historical Dictionary of Architecture *(Papadakis, 1999).*

* * * *

Either we consider that the Good, the Beautiful, the Human, which have been built slowly, are worth defending, or we consider that history must be erased with the stroke of a pen and begin again from zero.

—JACQUES ELLUL, *THE EMPIRE OF NONSENSE*

IN THE FACE OF political vices and vicissitudes, public intellectuals like Julien Benda (1867–1956), Noam Chomsky (1928) and Edward Said (1935–2003), lamented what they invariably called the betrayal of public intellectuals. The betrayal itself concerned many a public intellectual's unwillingness or incapacity to speak truth to power, sacrificing moral integrity and moral courage in favor of promotions and monetary gains, coveted awards, book or television contracts, tenured or endowed professorships, or general fame and social acceptance. Treasonous intellectuals know how to avoid incisive moral stances in favor, at times, of uttering soothing platitudes, and at other times, of promoting outright lies that serve a ruling ideology or a political party.

By contrast, good public intellectuals not only honored the good and the just and challenged harmful orthodoxies, but they also had the moral courage to defend the oppressed or to expose the crimes committed by the state in the name of its citizenry, even to the detriment of their own physical safety, their positions in society, or their monetary remuneration. Good public intellectuals may not have regularly succeeded in helping to achieve civil liberties or in preventing aggression between nations, but by virtue of their stances for peaceful coexistence and for the universality of human rights, they became exemplars to be emulated by their fellow citizens. Quite importantly, good public intellectuals were intellectual rebels long before they were scapegoated or morally banished by the power élites and by those who conform themselves to the power of these élites.

Jacques Ellul (1912–1994) was one such intellectual rebel in a life filled with intellectual engagements with subjects such as technique, law, theology, politics, morals, and art. In a sequence of six books published between the early 1960s until the early 1980s, he painstakingly analyzed the workings of political life dealing largely with Western countries while also discussing political and economic conditions in the so-called Third World.[1] He exposed the palliative appearances of political life that veil the veritable dystopia resulting from the confluence of technological determinism, political illusions, state and institutional propaganda. Furthermore, Ellul showed the extent to which this deleterious confluence was independent from essential ethical concerns.

With the exception of few authors such as Edward Bernays, who aggressively promoted the most controlling kind of propaganda aimed at what he called the desire-driven masses, Ellul was preceded and succeeded by many studies denouncing the uses and abuses of state and private media. These studies showed the extent to which the masses were manipulated to acquiesce to certain sociopolitical programs or to induce the consumption of industrial products in the actual Pavlovian sense of conditioned reflexes.[2] Hannah Arendt, in the *Human Condition*, drew attention to the usually overlooked fact that consumption implies eating or ingurgitating, where culture

---

1. See the bibliography below.

2. Three of the most important studies on propaganda to appear around the Second World War in France were produced by Serge Tchakhotine, Jean-Marie Domenach, and Guy Durandin. See the bibliography below.

itself was to be consumed in the most metabolic meaning of the word. By multiplying the artifice of false needs, the society of consumption not only separated the human realm from nature but it inexorably situated the human within the realm of propaganda and consumption.

In 1939, biologist and sociologist Serge Tchakhotine (1883–1973) released an important study in social psychology: *Le viol des foules par la propagande politique*, (The rape of the masses by political propaganda). The book was immediately banned in France, ostensibly in order to appease the German government against whom the book was directed. Republished in 1952, the book analyzed what Tchakhotine called the "psychic rape" that was perpetrated by the Nazi propaganda machine under deceitful and skillful management of Joseph Goebbels. Tchakhotine, who was a friend of Ivan Pavlov, detailed four elements of manipulating social psychology: aggression, the immediacy of material interest, sexual desire and attraction, and the need for social norms to convey a sense of security. The book also underlined the intimate association between the mechanisms by which the masses were manipulated in relation to the formation of political beliefs and eventually to political actions according to the direction already set by the propagandists. Once this strategy is set in place, the public is easily manipulated much to the detriment of democratic institutions and policies. Tchakhotine also extended his analyses, in the 1952 edition, to the ideological propaganda within the Soviet Union as well as the beginnings of the Cold War.

In 1959, French intellectual Jean-Marie Domenach (1922–1997), published *La propagande politique* in which he noted the commonalities held by various political propagandists, to wit: the simplistic reduction of social ills to one single enemy and the orchestration of adverse action against this enemy, or scapegoat, while conducting a grossly exaggerated political discourse intended to have a multiplying effect in a manner similar to a contagious infection. But in his belief that poisons have antidotes, Domenach suggested a counter-project to propaganda, a counter-propaganda that could redress the lies of the first propaganda while using the same tactics, a *tactica adversa* that has increasingly pervaded international relations since the rise of the Cold War. French psychologist Guy Durandin (1916–2015), in a 1982 study entitled *Mensonges en propagande et en publicité* (Lies in propaganda and in advertising), underlined the deep psychological link between political propaganda and commercial publicity. Citizens of the consumer society were compelled or rather determined to action by two main factors. On the one hand, their desire to possess and identify with the products that proliferate in this society; on the other, their decisions, whether rational or emotional, are strongly influenced by the information at their disposal, but this information was made available to them, or rather imposed on them, by corporate advertising empires.

As these and other studies have demonstrated, propaganda and counter-propaganda, disinformation and counter-disinformation, seek not only to convince but rather to provoke or trigger certain actions, whether these actions are meant to elect a political candidate, or market automobiles. Domenach had already shown that the phenomenon of propaganda, on an essential level and despite general appearances, was similar in the USSR, China, or the United States. Because propaganda forms, or rather conforms, public opinion it aims not only at explaining and validating decisions made by the state and powerful corporations, but it also aims at structuring and justifying future domestic and foreign policies. Convincing the citizenry to accept wholly unfair health insurance practices while presenting them as beneficial to an abstract well-being is an example of domestic policy. Justifying the invasion of a sovereign country as national self-defense by saturating the mass media with false claims of imminent threats emanating from that country is an example of foreign action.

Propaganda and Political Illusion

Consequently, Ellul asserts in *Propagandes* (note that he uses the word in the plural) that propaganda is in essence totalitarian irrespective of the forms it might take. It is not solely relegated to totalitarian states. The democratic state needs propaganda in order to compel citizens to embrace the fact that their society is technologically determined. Information, the press and publicity are included under the larger rubric of state propaganda; but above all education has been transformed into the most efficient model to enforce conformity to the technological society.[3]

Ellul also warned against what he called the "propaganda of integration" especially when the state's institutions and media are turned toward the state's own citizenry.[4] Propaganda is needed by the state to centralize its power and to disseminate the appearance of such centralization, giving the impression of strength and stability to friend and foe alike. What distinguishes the modern state, in comparison to all states in history, is that whereas they have all relied on propaganda the modern state practices propaganda as a method of governing. The political illusion is therefore, to a considerable degree, a result of propaganda. Moreover, propaganda and illusion have become inseparable parts of how the state justifies itself,

3. Ellul, *Propagandes*, 126.
4. Ellul, *Propagandes*, 9.

and how it claims certain political or ethnic identities or a mixture of both. Ellul concluded that propaganda deeply transforms the state's public and private realms. Its omnipresence produces a highly exaggerated or inflated political discourse becoming, in Ellul's words, a "machine to manufacture conformism," or to "manufacture consent" in the words of Edward Hermann (1925) and Noam Chomsky.[5]

The very consciousness of social classes, of national identity, of a certain "way of life," are shaped, enforced, and even exasperated by social propaganda and political propaganda in particular when it comes to conflicts between groups. But Ellul did not consider the victims of social and political propaganda to be necessarily innocent, not completely. They not only accept the contradictions of their condition, but in expressing their own engagements or opinions they are actually repeating what the propagandized discourse had already dictated. Ellul concludes, that the deeply alienating aspects of the technological society become readily accepted and justified by a citizenry that has been conditioned to rely on propaganda in order to justify a way of living. Citizens and politicians alike have repeatedly fallen into a great illusion by imagining that the state can truly be directed or its policies modified according to the citizens' principled wishes, and by naïvely believing that their participation in orchestrated electoral derbies is somehow a proof and a guarantee of democratic life. Ellul's notion of *l'illusion politique* can very well be termed *l'affabulation politique*—a political fairy tale.

Increasingly today, politics rests on the appearance of public participation and consent—an appearance worn by politicians who intend on propagating it to as wide an audience as possible. Ethical concerns, we are frequently told, are dialectically shared by the public and the politicians who claim to represent them. Yet public opinion drowns in an enveloping psychological miasma of words and images created and propagated by the empires of mass media, and one of the most insidious parts of this miasma is the phenomenon of the sound bite.

Sound bites have become the expression of choice and the weapon of choice in political exchanges in government and commercial corporations, or in the deliberations of universities that have become corporations. Sound bites, one needs to recall, are an amalgamation of words that have been radically transformed by the power of technological images[6], in particular those of television, and designed to deliver potent emotional charges for the

---

5. Hermann and Chomsky, *Manufacturing Consent*.

6. On the technological, or *technicist*, images in the realm of art, see Ellul, *L'Empire du non-sens*.

purpose of inciting into action or adhering to a vaguely articulated policy. Sound bites and advertising are quintessential products of the technological mindset, hence their joint deployment in order to augment consumption or frame a political discourse.[7]

One irony anent advertising empires is the selling of politicians, educational strategies, motion pictures, or vegetable soup with the same exacting efficiency, notwithstanding their true content or quality. One irony anent the sound bite is its deceptive brevity and superficiality in comparison to the enormity of social challenges which are then covered with an inflated discourse. The twitter phenomenon is but a continuation of the earlier phenomenon of the sound bite, and public intellectuals would be mistaken to dismiss it as a passing fad because it has become a preferred way of expression for adolescents, for politicians steeped in polished advertising strategies, and for a president of the United States, where it is increasingly clear that the brevity and superficiality of their tweets are but a reflection of the brevity, superficiality, and ineptitude of their thinking.

Revolution and Revolt

Given the deep and conjugated effects of the aforementioned phenomena, in particular their reduction of human freedom, one wonders where are the revolutions and why do they not multiply? In fact, Ellul dedicated no less than three books to the possibility of revolutions and the impediments to their realization (*Autopsie de la révolution*, *De la révolution aux révoltes*, and *Changer de révolution*). In *Autopsie de la révolution*, published shortly after the events of May 1968 when talk of revolution filled the air and student revolts occupied the center stage of political life, he drew attention to the difference between revolts and revolutions. A revolt is an uprising, an unrehearsed expression of human will based on a visceral dissatisfaction with a given status quo, with the social injustice of certain politico-economic conditions resulting in disenfranchised social classes.

Revolts may depend on or use a scapegoat (whether an individual or a group) to arouse their activities, and here Ellul indirectly joins French philosopher René Girard in emphasizing the role of the scapegoat whose very victimhood and innocence serve to balance the conflicts between antagonistic groups. But for all their intentions for immediate change, and despite the fact that they may be led by intellectuals who understand the happenings of the time, some revolts may not necessarily articulate a viable reformatory program.

7. See Scheuer, *Sound Bite Society*.

Other revolts fail to enact their reformatory programs because of the strong and violent opposition posed by a recalcitrant establishment. That is why, for example, the Arab Spring (2010), beginning in Tunisia, then Egypt, and spreading to other countries, the Spanish indignants (2011), and the Occupy Movement in the United States and other countries (2011), Black Lives Matter (2013), Standing Rock (2016) and others, remained on the level of uprisings or incomplete revolts because they did not succeed in changing the status quo and thus arrive at the stage of revolution. It is crucial to note, especially with respect to Egypt and the United States, that both states unleashed their full policing power in order to infiltrate, discredit, disband, and ultimately destroy these uprisings with an iron fist. Who can forget the armed forces, including F15 fighter-jets, that descended on Tahrir Square in Cairo or the militarized police attacks on Zucotti Park in New York, on Black Lives Matter protesters in Fergusson, and on the Water Protectors at the Standing Rock Indian Reservation?

It is also important to note the considerable role played by the corporate media in discrediting these reformatory voices, at times by accusing them of not knowing what they want even if their demands were remarkably precise contrary to the empty and usually unfulfilled promises of politicians, and at other times by engaging in what is journalistically known as the character assassination of their leaders. It is also vital to mention the names of the few courageous public intellectuals who not only wrote and lectured but also descended into the streets and conducted teach-in sessions, e.g. Noam Chomsky, Cornell West, and Chris Hedges.[8]

What is in common between revolts and revolutions is seeking a decisive break with deleterious conditions or chain of events in favor of a new departure, a better starting point. But revolutions differ from revolts in the sense that revolutions are based on an already elaborated ideology that revolutionaries seek to realize by overturning previous institutions. Once institutions are overturned revolutions install themselves as the new institutions at times fulfilling and at times negating or even betraying the earlier principled impulses and programs.[9] But Ellul's reading of Marx gave him a long-term historical perspective that made him see revolutions as a normal part of social evolution with the aim of arriving at a more just future order.

Talk of revolution has now almost died down in societies that are under the pervasive influence of technological-consumer society where consumers' manufactured needs are met to an excessive degree. Ellul lamented the

---

8. The teach-ins of the Occupy Movement have a very effective precedent in the teach-ins on U.S. campuses in the mid-1960s to oppose the Vietnam war and promote equal civil rights.

9. Ellul, *Autopsie de la révolution*, 137.

dulling effect of propaganda and the politics of illusion on the very senses of would-be revolutionaries while openly wishing, at the same time, for a revolution to oppose the technological society because of its restrictions on human freedom. Quite significantly he worried that the technological society has at times replaced human freedom, and concluded that the era of revolutions is gone, while still entertaining the hope that a revolt can be had, and perhaps even a remote possibility of revolution.[10] Any change now must come at the expense of immense efforts because the structures of technological society are entrenched, interdependent, and irremediably entwined with the power and modes of operation of the state.

Possible Revolution?

In *Changer de révolution*, (the last of the six books), Ellul recalled Marx's warning about the internal contradictions of capitalism, more specifically the concentration of wealth and power in the hands of an ever smaller number of individuals at the expense of an ever increasing number of exploited citizenry at home and abroad. His doubts about the possibility of revolution notwithstanding, he suggested that rather than taking hold of political power would-be revolutionaries could seek to harness the "positive potentialities of modern techniques and to orient them uniquely to human liberation."[11] Achieving this re-orientation requires nothing short of liberating society from its politics of illusion in order to reconstruct politics on the basis of humanist values.

Here, several contradictions appear in Ellul's thought, because of his repeated assertions in multiple publications of the autonomy of technique from human feedback, of the nearly inseparable relationship between technological society and the state, of the overwhelming power of the politics of illusion, of the ills of the society of spectacle, etc.. Scholars and critics of his work have noted these contradictions, but rather that expecting the consistency of an unchanging mind (a rare occurrence) is it not more reasonable to assume that his thought simply evolved over the years in engagement with the lessons of daily political experience?

But Ellul made another volte-face in *Changer de révolution* and proposed a far-reaching global reform, which is nothing short of a worldwide revolution/evolution in order to achieve right human relations.[12] His reform covered two scales: the national and the international. It was aimed at the

10. Ellul, *De la révolution aux révoltes*, 373–78.
11. Ellul, *Changer de Révolution*, 256.
12. Ellul, *Changer de Révolution*, 245–56.

technological society within the so-called First World as well as its relations with the so-called Third World.

On a national scale it consisted in:

1. Encouraging cultural diversity in all societies;
2. Reducing the daily working hours;
3. Eliminating salaries and replacing them with an equitable sharing of resources amongst all citizens.

On an international scale the reform consisted in:

1. Reorienting some of the West's productive capacities toward sharing its immense wealth with the Third world in order to meet its needs for optimal human development;
2. Renouncing the Western states' centralization and military empires.

Ellul proposed this reform even if, as he confessed, he was not necessarily convinced of its chances of success. But then this brief optimism changed considerably only a few years later with the release of his *Le bluff technologique* (*The Technological Bluff*) where he despairingly stated that he presently believed that "the game is lost and that the technological system, exalted as it is by the power of information, has escaped definitively from the directional will of humanity."[13]

This chilling statement, was based on observing deleterious social conditions, mediated by the technological system, as of the late 1980s, but these conditions and many more have become exponentially worse, in particular in the United States. Giant corporations in the finance, petroleum, pharmaceutical, and insurance industries benefit enormously from the Federal Budget despite their massive failures resulting from what has rightly been known as the "casino economy" and their threat to international economies as well as the lives of countless millions. The U.S. Supreme Court allows elections to be bought by authorizing unlimited corporate contributions to state and national elections under the guise of "freedom of speech." Education is being privatized and the arts and humanities have been commodified. Students who are burdened by heavy loans face diminishing chances of being able repay them even if they declare bankruptcy with all its incident corollaries. Labor unions have been crushed by successive government policies that favor corporations while wages stagnate and unemployment rises, and while chronically unemployed citizens are no longer counted in government statistics. Foreclosures and bankruptcies abound while social

---

13. Ellul, *Le bluff technologique*, 203.

safety services are subjected to a frontal assault in the name of economic austerity. Government domestic spying records and stores the electronic communications, telephone calls, as well as the financial and medical records of the citizenry. Indeed, the entire world is being watched openly and unapologetically. Government rendition, torture, and assassination programs, conducted in multiple facilities around the world, are frequently provided cover and justification by a media that serves corporate sponsors who benefit from massive government contracts, in particular the arms manufacturing corporations. Retired military officers engage in a revolving door policy between intelligence institutions (both public and private) and the media where they are usually called, in their roles as paid experts, to justify any given military intervention.

Many more examples can be added to this brief list, most importantly the ravaged ecosystem, but it is crucial for citizens to realize that these profoundly deleterious developments operate within a much-heralded participatory democracy where they are called upon to vote for a set of choices that have already been determined. The most superficial democratic appearances and processes are kept alive in the corporate state in order for citizens to be convinced of their political power, which, in the final analysis, has been declining rapidly. The fact that citizens choose to believe that they are living in a democracy—a plutocracy in truth—while most of their choices have been predetermined and while their movements, readings, purchases, and habits are under constant surveillance, *that is the great illusion*. How could citizens vote against the interests of multinational corporations if these very corporations purchase elections and write the trade and tax laws that perpetuate their hold on political and economic power? If a hypothetical future revolution is to have any salutary effect, its power will have to be proportionately calibrated to oppose the formidable power wielded by the entrenched forces of the technological-consumer society and its trade in political illusions.

It is unlikely, in the short term, for these entrenched forces to be persuaded by reason and goodwill alone to adopt a program such as Ellul's five-point reform. The price of resisting, revolting, or undertaking a veritable non-violent revolution will certainly be quite high. If only more intellectuals would descend in the streets and join hands with indebted students, underemployed teachers, unemployed plumbers, inactive brick masons, hungry musicians, and the forgotten homeless!

More than ever now, to revolt is a moral imperative because justice is a moral imperative. The courage of moral reliability depends on those who refuse to collaborate with the pitiless destruction of culture, as Hannah Arendt intimated in *The Origins of Totalitarianism*. Of the many lessons that the Arab Spring, the Occupy Movement, the Spanish indignants, Black Lives Matter,

Standing Rock, and others have shown, one deserves particular attention: by refusing the status quo and their insistence on genuine social reforms, these revolts frightened the corporate state which then proceeded to punish them with its militarized strength. The corporate state knew fear.

Bibliography

Domenach, Jean-Marie. *La propagande politique*. Paris: Presses universitaires de France, 1950.

Durandin, Guy. *Les mensonges en propagande et en publicité*. Paris: Presses universitaires de France, 1982.

Ellul, Jacques. *Autopsie de la revolution*. Paris: Calmann-Lévy, 1969; *Autopsy of Revolution*. Translated by Patricia Wolf. New York: Knopf, 1971.

———. *Le bluff technologique*. Paris: Hachette, 1988; *The Technological Bluff*. Translated by Geoffrey W. Bromiley. Grand Rapids: Eerdmans, 1990.

———. *Changer de révolution. L'inéluctable proletariat*. Paris: Seuil, 1982.

———. *De la révolution aux révoltes*. Paris: Calmann-Lévy, 1972.

———. *L'Empire du non-sens*. Paris: Presses universitaires de France, 1980; *The Empire of Nonsense*. Translated by David Lovekin and Michael Johnson. Berkshire, UK: Papadakis, 2014.

———. *Histoire de la propaganda*. Paris: Presses universitaires de France, 1967.

———. *L'illusion politique*. Paris: La Table Ronde, 1964; *The Political Illusion*. Translated by Konrad Kellen. New York: Vintage, 1972.

———. *Propagandes*. Paris: Economica 1962; *Propaganda*. Translated by Konrad Kellen. New York: Knopf, 1965.

Hermann, Edward, and Noam Chomsky. *Manufacturing Consent. The Political Economy of the Mass Media*. New York: Pantheon, 1988.

Scheuer, Jeffrey. *The Sound Bite Society: Television and the American Mind*. London: Four Walls Eight Windows, 1999.

Tchakhotine, Serge. *Le viol des foules par la propaganda politique*. Paris: Gallimard, 1939.

9

Jacques Ellul, the Symbol, and the Political Illusion

BY David Lovekin

> *David Lovekin is Professor of Philosophy Emeritus at Hastings College, Hastings, Nebraska. His PhD is from the University of Texas at Dallas. He is the author of* Technique, Discourse, and Consciousness: An Introduction to the Philosophy of Jacques Ellul *and editor with Donald Phillip Verene of* Essays in Humanity and Technology. *He is co-translator of Jacques Ellul's* The Empire of Non-Sense: Art in the Technological Society. *He has published numerous essays on Ellul and Giambattista Vico that deal with technology as a problem for the philosophy of culture in the spirit of Ernst Cassirer with the idea of technology as a symbolic form. His most recent essay is "Looking and Seeing: The Play of Image and Word—the Wager of Art in the Technological Society." He is an* Ellul Forum *contributing editor and a member of the International Jacques Ellul Society board of directors. He was a recipient of five National Endowment for the Humanities Grants. His photography has been published in numerous books and periodicals; his most recent work appears on covers of the* Prairie Schooner, *the literary magazine of the University of Nebraska, Lincoln, and in the Modern Arts/Midwest/Midtown gallery in Omaha, Nebraska.*

* * * *

THE MORNING OF SEPTEMBER 11, 2001, I showered and then went downstairs to a living room, turned on the television, and saw towers smoldering. I saw these moments as images on television like other images—an advertisement, a film, and old documentary—and not as a terrorist attack. Later, on reflection, I saw these as images differently. Someone does not like us. Who is this someone, these someones? Many of us still wonder what this attack meant, wonder who planned and perpetrated it. Many no longer wonder, however. Wonder dies as the images of this disaster that killed 2,996 people are endlessly displayed and used to support political propaganda. This date and these images are displayed each year as moments that changed America—America the victim became America the aggressor. The attack was used to help promote, falsely, the invasion of Iraq. And it changed how America viewed the world and how the world viewed America. The world viewed—that's the key—the world as image, as the appearance of meaning. In his essay "On Terrorism, Violence, and War: Looking Back at 9/11," Patrick Troude-Chastenet, citing Baudrillard, states: "This terrorist violence is not 'real.' It is worse in a sense: it is symbolic."[1] At issue is the difference between symbols and images.

Jacques Ellul urged us to comprehend the effects of technology in ourselves. I did not see the burning towers as "real." I saw them as images, perhaps a film constructed to entertain and to aid the passing of time. I did not see them as symbolic, referring to anything beyond themselves. At this time in 2017 they are often used to inflame, to tell us we are in danger from terrorism. They are, in Ellul's sense, clichés, propaganda urging support of our government and/or the mass media, if they can be easily separated.[2] They are, I contend, no longer symbolic, in Ellul's sense, but are part of the political illusion.[3] They are used to inflame but also to unite us in the project technology has become.

The symbol is a unity of oppositions that engages us dialectically, Ellul would say. The towers are symbolic when we wonder why they were attacked and by whom. They are symbolic when we consider that the terrorism they engendered would have been impossible without technology, the airplanes and the media that recorded them. They become symbolic when we see them as a terrorist agenda to reduce the lives of many Americans to exemplars of Western decadence. They certainly become symbolic when we ask about the nature of the symbol and of the images symbols have become. Now, the following questions apply: how does Ellul understand the symbolic in relation to the mentality that is technology; how does the symbolic play into Ellul's theological perspective; and how do the theological-sociological analyses stand with the symbolic in their relation to the political?

1. Troude-Chastenet, *Ellul Forum*, 57.

2. See my discussion of Ellul's notion of clichés as a form of consciousness in Lovekin, *Technique, Discourse, and Consciousness*, 201–7.

3. Lovekin, *Technique, Discourse, and Consciousness*, 191–95.

Ellul claims that politics has become an illusion in relation to the fate of the symbol. The symbols of myth and religion enabled pre-technological societies to form in meaningful ways that became narratives by which people could live. The Bible is one such narrative. *Technique*, however, is a mentality at war with the symbol that it reduces to the abstractions of logic and science. When technique/technology becomes the autonomous new sacred, symbols collapse and myths become false stories, clichés and mass media distractions. Values, ideals, ethos and character are expressed in the gasp: *n'importe quoi* or "Whatever."

The Symbol as Ground of Meaning and Boundary

In "Symbolic Function, Technology and Society" (1979) Ellul wrote:

> Man cannot have a relationship with another save by the intermediary of *symbolization*. Without mediating symbols, he would invariably be destroyed by raw physical contact alone. The "other" is always the enemy, the menace. The "other" represents an invasion of the personal world, unless, or until, the relationship is normalized through symbolization. Very concretely, to speak the same language is to recognize the "other" has entered into the common interpretive universe; to display recognizable or identical tattoos, for example, is an expression of the same universe of discourse.[4]

The symbol presents a mid-world between subject and object as "others" in separation. The human does not live in a world of sheer immediacy. These separations are taken up in a symbolic act that requires imagination and a sense of "otherness." The present is never merely present but is also a presence of absence or absences, of what could and should be. The notion of a world of simple facts is belied in the initial power of the symbol, where significance and meaning and the making of facts reside in a social collective memory. Ellul again:

> Consequently, that which will be retained in the collective memory is not the materiality of a fact, but its significance—its position in relation to the ensemble of symbols by which the group lives and the possibility of mastery over fact which is thus given to man. Hence collective memory assures the transmission of this universe of meaning, constantly enriched by new symbolic acquisitions, which permit new symbolic possibilities

4. Ellul, "Symbolic Function," 210.

that can only be constituted because they can be inserted into a previously existing system of symbolization.[5]

As a historian of law and social institutions Ellul had noted that the aristocracy in Rome was not formed by the facts of heroic acts but by narratives:

> [G]reat deeds were collected, transformed into an epic *account*, and then reconstructed in such a fashion as to become symbolic. At this moment, a double movement is produced: one moves towards the heights, further from the origins, as the eponymous ancestor becomes the concentration point of symbols, and is attached to a higher symbolic origin. This results in a god—goddess, or demigod—who is established symbolically as the true origin and as the explanation of the progenitive power of the ancestor.[6]

Symbols establish boundaries in fields of significance grasped in relation to what is and is not located by a sense of origins. A person is great in relation to an ancestry beyond the present. A society gains a similar sense of origins and power in a sacred symbolically constituted in this double movement of consciousness between a present and a past.

In his *Histoire des institutions*, Ellul writes: "The Roman sacred is at the same time both religious and magical. It is religious in that it worships the transcendental powers and it is magical in that it utilizes those powers which are immanent."[7] Although Ellul is not always consistent with his use of "myth," mostly he means stories established symbolically in an immediacy and not established scientifically or technologically by reflection. Magic, the most primitive form of technique, depended upon myth, upon a qualifying and limiting sense of power beyond which the human must not go. This is also clearly adumbrated in the *The Technological Society*, where true technique separates from traditional notions of the sacred both in nature, myth, and religion to become the new sacred "freed" from the boundaries of the past, only to become a boundary without limit, a deeply paradoxical development in which the very possibilities of "symbol" and "myth" are called to question.[8] Technology claims a freedom from the very conditions that made

---

5. Ellul, "Symbolic Function," 211.
6. Ellul, "Symbolic Function," 212.
7. Ellul, *Histoire des institutions*, 1:220–21. My translation.
8. Ellul, *Technological Society*. On the issue of primitive technologies and magic especially see 23–29.

it possible and against which it constantly struggles. The other is continually heralded by the cry of *n'importe quoi*, or "whatever."[9]

Ellul's sociological analyses are centered in *The Technological Society*, *The Political Illusion*, *Propaganda*, and *The Technological System*. The notions of the symbol are mostly behind the scenes, except in *The Political Illusion*. Politics becomes an illusion when it is considered to be the most important aspect of modern culture at the moment when politics no longer exists. It has become both the "necessary" and the "ephemeral."[10] Politics depends upon technology, which is spread throughout by the mass media that makes the acceptance of technology and its necessity possible. Ellul plays on the distinction between *le politique* and *la politique*.[11] The former addresses the values like truth, beauty, and goodness that transcend daily life; the latter adopt the concerns of method and implementation of efficiency, with values that, upon reflection, disappear in the process of technique that is manifest in the propaganda and clichés of the mass media. It has become an unquestioned absolute with no beyond. The beyond suggests the symbol that, in turn, requires language understood in a certain sense.

In *The Political Illusion*, Ellul notes that *le politique* requires a dialogic language not caught up in the ephemera of mere fact, the fodder of propaganda, if it is to effectively unify rather than inflame. Ellul remarks:

> Language reminds us of the necessity of tension at two levels: tension or contradiction is based on a similarity between signifier and the thing signified (when that tension disappears, there is no more language—that is why, whatever one may think, imagined reproduction of reality is not language); the other aspect is the tension between two interlocutors: if a difference does not exit, if they are identical, there would be no language because it would have no content; if a common measure did not exist, there would be no language because if would have no form.[12]

Images pervade the technological society and the mass media with a kind of necessity. They obviate the needs and abilities of language in the above sense. Words that refer, the vehicles of ordinary language, are "humiliated" in their reduction to images, to the idea that an image is worth a thousand words. Instead, an image may require a thousand words.

9. This notion is much developed in Ellul, *Empire of Non-Sense*. See 59–60 for this notion that is then further developed throughout the book. Thanks to Samir Younes for his good advice and insights.
10. Ellul, *Political Illusion*, 25–67.
11. Ellul, *Political Illusion*, 18n3.
12. Ellul, *Political Illusion*, 215.

Ellul notes in *The Humiliation of the Word* that a sound behind forces a turn of the head with an almost gravitational necessity to fix the sound in sight. Sight and sound are connected but portend different dimensions. The image and the word signal the two directions language and meaning take. The image is what is before me as appearance and is what Ellul calls the "real."[13] The word as sound surrounds and opens to the "true." The world of sight ushers in the logical and the world of sound brings the story, the narrative, symbols and metaphors.[14] Ellul states:

> The visual reality is clearly non-contradictory. You can say that a piece of paper is both red and blue at the same time. But you cannot *see* it as both at the same time. It is either one or the other. . . . Declaring that two opinions cannot both be true when one denies what the other affirms has to do with vision, which involves instantaneousness. But language involves duration. Consequently, what is visual cannot be dialectical. . . . Only thought based on language can be dialectical.[15]

I cannot see my friend as both friend and enemy although he may be so. In the domain of Aristotelian logic, A equals A; A cannot be both A and not-A. But in the domain of the word, contradictions may abide and meaningfully so. It is meaningful in the Christian tradition that God is three-in-one; it is meaningful in Homer's *Odyssey* that Athena can be both a god and a bird; it is meaningful for T. S. Eliot, that the sky can be an etherized patient. It is meaningful that enemies may also be friends. The tension between the image and the word—their presences as "others"—produces the symbol. They mediate what may be problematic in human experience like the divine in nature, the divine beyond human experience, and alienation in human experience. Symbolic mediation does not cancel oppositions but clarifies and situates them. It is now meaningful that the endless showings of the burning towers of 9/11 are not symbolic, that their meaning must be found in the domain of the word.

Ellul's discussion of the symbol suggests three moments: awareness, awareness of awareness, and then an awareness of that. There are no resolutions between those experiences. The technique of politics is to declare war on terror—war on an idea—that would be a war on itself. As said earlier, 9/11 required airplanes, skyscrapers, and the media to "immortalize" them and to turn them into clichés, substitutes for meaning, the fate of many words that

---

13. Ellul, *Humiliation*, 26–32.
14. Ellul, *Humiliation*, 19.
15. Ellul, *Humiliation*, 11n3.

began as symbols and metaphors. Technique continues to remain the problem that it cannot solve because it does not and cannot know itself.

According to the *Oxford English Dictionary*, "cliché" first appeared in 1832 and referred to a stereotype block, a printer's cast or "dab." It began in a visual dimension, but was also a variant of *cliquer*, to click, possibly the sound the striking of the lead pieces made. The auditory sense is lost in its modern sense as is the history and importance of the word. The word was a metaphor or symbol but is now only a worn-out expression. This sense is either unknown or forgotten as a sense of origins is denigrated. The word symbol, from the ancient Greek, suggests a throwing, the establishment of a contract, in short, the putting of things together through an action. Later, in old English, it connotes a theological doctrine like the Apostles Creed. And then in eighteenth century English it can refer to a chemical element or a mathematical quantity. The development of the logic of technique moving away from the symbol toward the concept and the cliché is witnessed in language.

The Collapse of the Symbol and the Humiliation of Meaning

The word and the image, as stated, indicate two different domains—the real and the true. The true and the real can only be double-edged requiring then the third level of awareness of awareness of awareness. Ellul wrote:

> [Language] can command an action. It gives birth to institutions. But reality is not where its specificity lies. We have mentioned myths and symbols, allegories and metaphors, analogies and history, as spheres in which language moves about easily. In these contexts it takes on its full stature and becomes truly the word. In other words, it is true to itself when it refers to Truth instead of Reality.[16]

Important then to face the truth of technique as it appears in reality and in language or in its absence.

The logos of technique demands the reduction of meaning to methods applied with a mathematics-like precision in search of the efficient as absolute in the embodiment of the "one best way."[17] Ellul is clear. Technique is an intention that transforms the "natural" object as other. In this way, like the symbol, technique initially occupies a mid-ground, but unlike the symbol that difference or otherness is canceled. Any technical operation using tools

---

16. Ellul, *Humiliation*, 22.
17. Ellul, *Technological Society*, xxv.

extending from mind and body like hammering, chipping flint, or programing a computer is put to question in a search for perfection by reason and rational judgment and is transformed into a technical phenomenon. The body and natural objects are the most immediate others. Traditional or relative means are not valued as social otherness is contended: nothing short of absolute efficiency or efficiency as an absolute will do. But what values does efficiency portend? For example, are more efficient means of killing, of poisoning, of deceiving good? Efficiency on this kind of reflection begs the question of choice in relation to value. Efficiency *per se* becomes nothing *per se*. Efficiency, not in the physicist's sense of output exceeding input, has become a cliché, a word to be repeated that refers to nothing beyond itself, and the cliché becomes an important technical phenomenon.[18]

By my count there are seven characteristics of Ellul's *technique*: rationality, artificiality, automatism, self-augmentation, monism, universalism, and autonomy.[19] Rationality is the key. Ellul states:

> This rationality, best exemplified in norms, and the like, involves two distinct phases: first, the use of "discourse" in every operation [under the two aspects this term can take (on the one hand, the intervention of intentional reflection, and, on the other hand, the intervention of means from one term to the other.)]; this excludes spontaneity and personal creativity. Second, there is the reduction of method to its logical dimension alone. Every intervention of technique is, in effect, a reduction of facts, forces, phenomena, means and instruments to the schema or logic.[20]

The discourse applied is the discourse of the image, which replaces the action or the object to produce the apparent true (the contradiction not allowed) as it is. I have added the clause in brackets that is missing from Wilkinson's otherwise fine translation. That clause emphasizes the double nature of awareness that is canceled in the use of "discourse" instead of furthered by it as in traditional language. The current cliché, "It is what it is," is telling, as if something is merely A, which it would be without time or place. For the logic of the image, "is" is always "is." The truth, however, is that one "is" is to the left of the other "is" as I read it and also as I say it, but this conundrum is not addressed by the logic of technique. These words would

---

18. Ellul discusses the technical phenomenon as it arises from the technical operation in *Technological Society*, 19–22. See a more complete account of this in Lovekin, *Technique*, 52–187.

19. Ellul, *Technological Society*, 78–147. See my discussion of this in Lovekin, *Technique*, 152–87.

20. Ellul, *Technological Society*, 78–79. I have amended Wilkinson's translation with a phrase in brackets that he left out. See Ellul, *La Technique ou l'enjeu du siècle*, 74.

be replaced by more technical phenomena. As the hammer replaces the rock, so the hammer is replaced by whatever new version of perfection can be accomplished. Perfection is both present and always around the corner. To produce the thing in itself is the goal of technology in what has become our current metaphysics.[21] We, have, then, digital watch fashion, a never-ending series of "nows" or "not yets," of awareness and further awareness but no awareness of these awarenesses.

Ellul wrote:

> The technological system performs unintentionally. Hence, wherever it is applied it produces a new kind of objectification which has nothing to do with Hegel's: it is no longer an objectification of the subject, and does not enter a subject-object dialectics. Now, anything that is incorporated or seized, is treated as an object by the active system. . . . Nothing can have an intrinsic sense; it is given meaning only by technological application.[22]

The technical phenomenon is, by definition, artificial: it is a construct of reason, understanding, and practice that denies its artificial state by forgetting or not considering the origins that made it possible. The natural other was not made but simply transformed. Concepts are, then, turned into objects and objects are conceptualized. To be is to be a label, for example; "Coca Cola is the real thing." The list is endless by definition.

I need to be clear: concepts are not symbols in that differences are subsumed under a generalization. I can develop a concept of "cat" as a genus, *felis*, and to a species, *felis sylvestris* or *felis catus* but none of these are my cat. No concept of a cat can produce a cat. They simply give us another dimension, but it is the conceptual dimension that technique requires with the elimination of otherness and difference, the things that Hero, my cat, embodies. The concept is a warehouse for particulars, useful for grouping or understanding, that can at any moment can become a concentration camp. As noted above, embodiment is cancelled by the technical phenomenon and with it so is the possibility of the symbol.

Technical choices are made automatically for the absolute of efficiency. Two is greater and better than one, etc. As one of our city entrepreneurs said in a side-splitting tautology: "more is more." Production is geometrical rather than arithmetic, self-augmenting in Ellul's terminology. Medicines and clinics can produce diseases as well as cure them. One car produces a road that requires gas that engenders industries galore, including insurance and fast food. No end is in sight. As a consequence, what can be done

21. Ellul, *Technological System*, 74.
22. Ellul, *Technological System*, 9.

will be done; no countervailing attitude is possible in the monistic pose. No weapon is made that is not used (any exception would prove the rule). Further, geography and culture provide no cultural boundaries. Cell phones and televisions play out in the most remote villages.[23]

Technology as the sacred, Ellul states, deserving the deepest respect, ironically appears as an image, and then as another image, *ad infinitum* and *ad nauseum*, before which we bow. As autonomous, technical subjectivity and objectivity form an uneasy identity. Technology becomes autonomous as the Wholly Other. But here is the problem: absolute meaning cannot be series of moments; meaning has to occur outside the series, any series. An infinite series cannot just be another element, which could not define the series. The summation of the series would result in a contradiction, which technique does not allow. Also, the notion of the infinite cannot be merely outside, an empty class concept and, therefore, truly meaningless. A true infinite must be both imminent and transcendent. Göedel's proof is one fine example.[24] For technique, what Hegel would have called a bad infinity, meaning is reduced to the here and now as a profusion of moments; the sacred that traditionally appeared in nature and in religious worship supporting a transcendental dimension is relocated.[25] Technology becomes the new sacred, as imminent as time measured by a clock. The proffered freedom is really an enslavement to a system created by humanity from which there is no escape because it is unknown like water to fish. The objects created by technology are no longer known as objects; natural and cultural objects are subsumed by a technical intention that vitiates all others. But, what do they finally mean apart from their co-option in the technical system?

Biblical Language and Symbol

Ellul's critique is incomplete without his theological perspective that faces of against the sociological. Here is where he finds his Wholly Other, in biblical literature and revelation. Ellul contends that human fate culminating in technique was revealed in biblical narratives that placed the Fall in the hands of those who refused God's guidance. After rejecting God, humanity leaves the Garden where all was whole to build cities and babbling towers in a garbled and incoherent language, not only unclear but meaningless.[26] These biblical stories embodied in a poetic history show a God revealed in

23. See my further discussion of this in Lovekin, *Technique*, 157–82.
24. Lovekin, *Technique*, 183.
25. On the Bad Infinity, see Lovekin, *Technique*, 98–105.
26. I am thinking here of Ellul, *Meaning of the City*, 19.

history by words that are also actions. God appears incognito in uncertain messages, which convey an obscure but crucial sense.[27]

In *The Politics of God and the Politics of Man* Ellul reads the Second Book of Kings to show that God's grace has nothing to do with human political power: the just and the unjust are rewarded or punished equally.[28] These narratives do not obey logical stricture but inform by contradictions that place meaning in symbols beyond ordinary logic where A must be A and ~ A. It is a logical impossibility for God to be Jesus, a man, and the Holy Spirit, and yet this metaphor offers the Christian hope for a reality beyond measure and finitude. God allows transgression; he sacrifices himself; he shows the power of inaction and forgiveness; he offers freedom unfettered by the desires of technique that promises but does not satisfy. Technique offers one thing after another with one as good as the other, an endless and unbridled infinity, what Hegel called a bad infinity. Peace is only found in God as the Wholly Other who was once a man, a self-generating boundary. An infinite series with no boundary can only end in contradiction, the type of meaning that technique abhors; contradictions must be resolved in absolute identities. Technique claims such a status, the mantle of absolute truth, paradoxically, with a denial of absolute truth, again that ends in contradiction intolerable for technique.

Technique marches unwittingly to the drums of irony. While claiming to desacralize culture, to rid it of the embarrassments of myth and superstition, it apes the logics of both. Myth and the sacred adorn the phenomena of technique as appearances, like the sunshade on McCormick's first reaper.[29] I contend that Ellul's biblical studies and his sociological studies come together, albeit dialectically, with the notion of the symbol. Ellul weighs technology's failures by the clichés and platitudes—denuded symbols—that decorate modern life with the mediocrities blathered endlessly by the "mass media." In short, symbols establish true boundaries, and thus they "refer," while clichés do not.

With the American flag we have a symbol that refers to the so-called founding fathers, to the Constitution, to the Declaration of Independence, to a tripartite form of government, to justice, etc. It offers freedom at a price,

---

27. In Ellul, *Hope in the Time of Abandonment*, God enters history incognito. Finding his presence in signs provides the mask of hope. See especially 192–222. Also, the importance of Kierkegaard should never be forgotten. The theological dialectic between the sign and symbol, the presence of absence, and the hidden God revealed in history warrant more discussion. A good place to begin is Van Vleet, *Dialectical Theology*, 16–20 and 64–76. I would like to thank him for his help with this essay.

28. Ellul, *Politics of God*, 33–36.

29 Ellul, *Technological Society*, 74.

at the presumption that all are taken as equal under the law given the realizations of what we can and can't do. The "All" should not be a night in which all cows are black. Laws require limits and precedent and interpretation and disagreement. The American flag, as a symbol, is constantly under threat as are its laws. We can reduce the flag to a hunk of cloth and make serious prohibitions against burning it, as if the cloth was freedom, was the country. At that point it would no longer refer in a tension between the symbol and the symbolized, which are not meaningful as identities. The flag is not the country, but when it becomes so, the true country vanishes. The second amendment does not mean guns; it refers to the forming of militias and thus has to be understood in an historical context in which guns may or may not be involved. In no sense is the arms industry mandated, although we have here a clear symptom of technique's grasp.

Technique as a mentality mandates the reduction of all to technical phenomena, in which logical concepts are embodied. The human body and other natural objects are dis-embodied and reduced to an ephemeral status, illusions projected by technique. The human self and the human body are crucial for understanding the symbol; in my view they make it possible, although I'm not sure Ellul would have agreed with me here. Nonetheless my views spring from his analysis. I could further conclude—again without his permission—that God, as Wholly Other, was killed by human sacrifice to the demands of hubris and busyness but then is resurrected, strangely by the means of murder that attack the body. God becomes man who sacrifices the man/God on a Roman device for capital punishment. The proliferation of devices from Jesus to the present is endless, truly a bad infinity. The technical phenomenon and its effects—the phony crosses we have to bear—are the means of disincorporation, the death of politics and the symbols that made it possible.

Conclusion: Symbol, Boundary, and Political Possibility

I regard the symbol as the link between Ellul's social and theological analyses. *Le symbole* and *le politique*, however, must be understood in process and context. First, symbols are the essential ingredients for natural language, for the means of communication with communal purposes. As I have indicated above, these symbols at first appear in those languages and in the myths and stories that illuminate and instruct those communities. They presume a separation of subject and object, person and world, that is united with and by the symbol. Language and some sense of the political are thereby coextensive in a move for a beyond, a horizon, and a possibility, the conditions of a meaningful freedom that connect subject with object. Perhaps as the ancient meaning of symbol as a throwing things together reminds, they suggest a sense of play, perhaps a contract for establishing play with values for the communal good as goal. Symbolic union, however, is provisional and requires an imaginative play outside the seriousness of rational discourse but play that is not to be taken lightly.

Further, the "other" is the symbolic aim for value and meaning beyond techniques. *Le symbol* is a reach and a stab for *le politique*. This reach is thwarted by *la technique* and *la politique* in its establishment of false identities. As Ellul wrote in *The Technological System*:

> The symbol in the technological system has changed meaning and value for the plain reason that the symbolized object or the object provoking the reference to symbols is not what it used to be. It is no longer an object both alien to man and belonging to a "natural" universe in which everything had to be symbolized. The object of the technological world now has its own efficiency, its power, it can obtain results, it is a work of man and yet alien. Hence, the symbol no longer plays the same role as earlier in regard to the object.[30]

The object has become the technical phenomenon and then the system of those phenomena that refers to nothing outside the system beyond the clichés that announce the symbol's humiliation and the end of play. Boundaries are effaced in the refrain of "whatever." Thus, as the symbol is effaced, technique itself cannot be symbolized from within the system.[31] Ellul's stance is from the standpoint of the symbol, his version of the wholly other in Christian form.

I have not had Ellul's conversion experience but, instead, have found meaning in his words, for example:

> The most explicit and the best-explained word still brings me inevitably back to mystery. This mystery has to do with the other person, whom I cannot fathom, and whose word provides me with an echo of his person. I perceive this echo, knowing that there is something more. This is the mystery I feel as I recognize spontaneously that I do not understand well or completely what the other person says. There is a mystery for me in my own lack of comprehension, as I become aware of it. How am I going to react? How can I respond? I sense a whole area of mystery in the fact that I am not very sure I understand correctly. I am not very sure about answering. I am not very sure of what I am saying.[32]

30. Ellul, *Technological System*, 176.
31. Ellul, *Empire of Non-Sense*, 66.
32. Ellul, *Humiliation*, 25.

Ellul's echo is not the one dogging Narcissus who has to learn that he is and is not his image. Ellul is what he is and is not—a Christian seeking a hidden God proclaimed in the Biblical text steeped in enigmatic ironies; his Bible is not a machine churning out proclamations and rules. He is not the dogmatic Christian who claims to know God's words directly but that are only known symbolically. Ellul's words open us to the problem of meaning without boundaries or limits that has attracted all the great thinkers from Plato to Whitehead. The sound of his words requires a turn of the head to a dimension that surrounds. We stand before his symbolization of technique viewed from beyond the "system" to wonder whether it is so or not.

Doubt is the handmaiden of thought that dissipates in certainties; imagination is the tool for finding the ground of meaning in the symbol, humiliated by the clichés of twenty-four hour news, by the prison houses of social media, and by Tweets and other barbarisms. We must look for the shades of a truth beyond the images that obfuscate the so-called political landscape, an absence of meaning that echoes truth nonetheless.

Bibliography

Chastenet, Patrick-Troude. *Ellul Forum* 57 (Spring 2016).
Ellul, Jacques. *The Empire of Non-Sense: Art in the Technological Society*. Translated by Michael Johnson and David Lovekin. Berkshire, UK: Papadakis, 2014.
———. *Histoire des institutions: L'Antiquité*. 2 vols. Paris: Presses Universitaires de France, 1992.
———. *Hope in the Time of Abandonment*. Translated by C. Edward Hopkin. New York: Seabury, 1973.
———. *The Humiliation of the Word*. Translated by Joyce Main Hanks. Grand Rapids: Eerdmans, 1985.
———. *The Meaning of the City*. Translated by Denis Pardee. Grand Rapids: Eerdmans, 1985.
———. *The Political Illusion*. Translated by Konrad Kellen. New York: Knopf, 1967.
———. *The Politics of God and the Politics of Man*. Translated by Geoffrey W. Bromiley. Grand Rapids: Eerdmans, 1972.
———. *Propaganda*. Translated by Konrad Kellen and Jean Lerner. New York: Knopf, 1965.
———. "Symbolic Function, Technology, and Society." *Journal of Social and Biological Structures* 3 (1978) 207–18.
———. *La Technique ou l'enjeu du siècle*. Paris: Armand Colin, 1954.
———. *The Technological Society*. Translated by John Wilkinson. New York: Knopf, 1964.
———. *The Technological System*. Translated by Joachim Neugroschel. New York: Continuum, 1980.
Lovekin, David. *Technique, Discourse, and Consciousness: An Introduction to the Philosophy of Jacques Ellul*. Bethlehem, PA: Lehigh University Press, 1991.
Van Vleet, Jacob E. *Dialectical Theology and Jacques Ellul: An Introductory Exposition*. Minneapolis: Fortress, 2014.

Part Two: Applications

The ten chapters in Part Two are a kind of smorgasbord of varied applications of Jacques Ellul's thought. The writers of these chapters are a diverse crowd from distinguished veteran professors to emerging young scholars and graduate students, from politician to medical doctor to librarian to community organizer. Some have studied Ellul for decades, others are newcomers to the Ellulian vineyard. Some draw deeply and widely from Ellul's writings, others sample more lightly. Some write from an obvious faith perspective, others are appropriately secular in approach. What is abundantly clear and on display is the incredible range and variety of impact Ellul has had and continues to have on so many.

In chapter 10, community organizer Sylvie Justome tells the story of how Ellul's ideas and values helped a small community organize to stop a dangerous chemical business project in a town near Bordeaux. Then (chapter 11) senior Ellul scholar Langdon Winner reflects on the political "revolution" rhetoric, especially during the sixties in America—and the reality and aftermath. Paul Heidebrecht (chapter 12) describes a fascinating political experience with the Ottawa government followed by (chapter 13) an interview of Mayor Robb Davis by Mark Baker: what is it like to be in political power while drawing deeply from Ellul's writings? Virginia Landgraf (chapter 14) uses Ellul to assess the recent political history of Thailand where she has lived and worked. Graduate student Zach Loeb (chapter 15) was involved in the 2011 Occupy Wall Street movement and recent PhD Carman Madorran (chapter 16) has been a close observer of new political movements in her native Spain (Podemos) and comparable movements in Greece and Italy. It is fascinating to get their Ellulian take on these attempts at protest and a new revolutionary politics. Dr. Raymond Downing (chapter 17) and doctoral student Jason Hudson (chapter 18) draw on Ellul for insights for the medicine and ecology arenas. Finally, graduate student Michael Morelli (chapter 19) draws from Ellul's work to comment on the rise of Charlottesville-style white supremacist protest and the challenge of finding a faithful response that addresses the true roots of the problem.

10

Resistance in Thought and Action
From the Global to the Local at Verdon

BY Sylvie Justome

> *Sylvie Justome holds a degree in classical literature from Paris-IV Sorbonne University (France) and she is agrégée. She has lived and taught in Egypt, Côte d'Ivoire, Mali, and Djibouti, where she supported the development of literature, theater, and cinema. She then became an inspector with the Académie-Inspecteur Pédagogique Régional and worked in the French overseas territories of Nouvelle-Calédonie, Polynésie française, and Wallis & Futuna. She is a longtime member of the Association Internationale Jacques Ellul and has promoted Ellul's thought within the association she created and leads, Agora Soulac Energie, and within Collège Jacques Ellul in Bordeaux.*

* * * *

FOR A FEW OF US activists, reading Jacques Ellul's work on *la technique* inspired not only much thought and reflection, but also a free, non-violent action from 2006 to 2009 to oppose the construction of a liquid gas tanker terminal near our community. It was David against Goliath. The odds were not in our favor; it was a very long shot. We were only a handful of local inhabitants directly threatened by this dangerous project. As soon as we began objecting we were accused of selfishness and a "NIMBY" ("Not In My Back Yard") attitude. We couldn't match the forces we were up against. We had to get as much press and attention as possible for our little micro project. Our strategy was to point out the planetary dimension of its consequences and emphasize the universality of what was at stake. In short, we were applying Ellul's principle "Think globally act locally."

Why did we do this? We rejected a materialistic, quantifying, mathematical approach to what is human. Every man, woman, and child is infinitely worthy and unique, no less important than a group. Human values and ethics can and should be upheld by each and every one of us as intrinsic and irreducible. This view of humanity gave us an incredible faith as we faced the coldly quantifying, mercantile Technique, Money, and Politics. It took us all the way to our victory, non-violent and illusion free. Of course, luck also played a part!

What follows here is a simple, first-hand testimony about our little victory, some oral evidence that the David against Goliath tale is more than just a fable! It is the account of a non-violent fight we won in 2009 against the power of out-of-control "Technique" in the hands of a multinational business called "4Gas" and its plan to build a liquid gas tanker terminal at Verdon, on the Médoc peninsula near Bordeaux in southwestern France.

The victory we won was inspired by, and illustrates, Jacques Ellul's thoughts about the greed and arrogance of modern technological "method," especially as he described it in *Le Bluff technologique* (the technological bluff). Ellul explains how the technical power believes itself to be omnipotent—and almost everybody accepts that claim:

> It's indeed everyone's core belief that technology can achieve all of our desires for power and that it is, by and in itself, all-powerful.... That is the absolute belief of our "modern" world which implies mankind's absolute renunciation to taming it. ... The mightier the power is, the harder it is to master.... Humanity gives up adopting reason as a critical power, precisely because reason seems to us to be concentrated in technical sciences and technologies.[1]

This account will show how we resisted and we'll explain how we won, because against the tools and method of "Technique" there remain some democratic tools with which to fight—and sometimes win.

In 1954 Jacques Ellul published *La Technique, ou l'enjeu du siècle* (literally "Technique, or, the stake of the century," or roughly "what's at stake this century"), translated and published in English as *The Technological Society* in 1964. That same year, 1954, Ellul also published *L'Homme et l'argent* (literally "Man/Person and money"), not translated and published in English until 1984 at *Money and Power*. Clearly Ellul was reflecting on Technique and on Money at the same time. We experienced the power of *both* Technique and Money first-hand during our own fight against the project to build the liquid gas tanker terminal in Verdon.

---

1. Ellul, *Bluff Technologique*, 295.

An international investment firm, hidden behind the façade of a pension fund, proposed the huge terminal project as a necessity for harbor development. For this they sought major investments. But, in fact, the announced development goals of the project didn't matter at all. The true objective was the just the collection of these investment funds to create short-term financial profits. Ellul was right on target in warning about both Technique and Money, by themselves, and all the more when linked together.

The first thing that should be said is how much the Médoc peninsula preserved its natural and quiet state over time. But Ellul often warned about how the technological method was being extended to all areas of the world and all aspects of life. Nowhere is anyone free from the influence of this new force. When I was working in the Polynesian islands, I already was aware that these peoples never asked for the French nuclear tests to take place in their archipelagos and yet they were and remain, terribly impacted by their effects. There is no clearer illustration of that truth than Verdon, this tiny place of nature, sun, sea, sand and trees, inhabited by just a small number people living year-round. Then suddenly we were all threatened by something huge and highly dangerous (classified "Seveso 2 high level" which is the very highest of risks): an industrial complex's technological vision and its accompanying financial schemes.

How did we feel when we first heard about this project (though it had been in the works for years without alerting those who were going to be affected)? Overwhelmed, of course! The majority were resigned and felt powerless. A few mayors in the region were seduced by financial promises into giving their support. But a very few of us felt angry and insulted. What did we do first?

We sought information, which was hard and exhausting to come by, for everything was carefully hidden behind reassuring and optimistic communication. "Don't worry, everything's under control by international experts." What about the elementary school within 800 meters of the site? "Don't worry." And how about several retired people who just invested all their savings in a nearby house? "Don't worry!"

But we were not about to give up. We researched the facts and scientific evidence. What exactly is "methane gas" (CH_4) for instance? We discovered that after emission CH_4 global warming power is forty-nine times higher than CO_2 after half a century, and still twenty-five times higher than CO_2 (3) after one hundred years. So the urgency was now, even more than in the long run. It was important to reduce CH_4 emissions without delay. And while some knew the truth about CO_2 and its greenhouse effects—- who knew about CH_4? We also learned that liquid natural gas is not actually "natural" but is compressed six hundred times with a high risk of explosion and ignition.

We also discovered that the Medoc inhabitants had spiritual brothers and sisters and similar minded groups elsewhere, facing the same challenges, for example, the Rabaska project in Canada. An Italian reporter met with us, shared a similar experience of resistance, and gave us some advice on how to communicate our concerns and opposition. We figured out that the newspapers *Le Journal du Médoc, Sud-Ouest*, and even *Le Monde*, only published information tightly controlled by the corporate publicists. Their in-house, so-called *"Impact Study"* clearly underplayed the risks to health, safety and security, environmental impact, and even to the local economy.

A kind of pseudo "public debate" took place from September to November in 2007. It was organized by the public authorities and consisted of a few meetings here and there. But we didn't really have a chance to speak and be heard there. The "debate" was a complete farce, a formal spectacle. While anybody could in advance send written questions, objections, or approvals, it was really just about creating an appearance, an image, exactly as Ellul wrote, to get everyone muddled and confused.

The methods they used were classic. Make any objectors feel guilty. "You should support the project because it is in the national interest for French energy independence!" (In reality, France was already over-equipped for stocking natural gas). "You should be ashamed to react with a selfish "NIMBY reflex!" Fear and intimidation were stoked up by having a meeting hall overflowing with workers from the Bordeaux harbor, brought there by special buses. Financial promises were made to win support from small town mayors in the region. External impacts were conveniently overlooked, such as the necessity of laying gas lines through the Médoc vineyards.

We also called our political/government representatives for help. But neither at the local level of the "department" (county) nor at the national level did we find anyone inclined to listen. Some of them gave us contradictory or unclear responses that went nowhere. Others, like Alain Juppé, the former French Prime Minister and Mayor of Bordeaux, were deeply in favor of the project and dismissed our concerns and protest as "hysterical." As Ellul said, politics is an illusion if we believe politicians will provide solutions to our essential problems. Not just the politicians but the intellectuals, such as Michel Serres, laughed at us and did not help. We also called on the media for help our cause and concerns, but, as I already said, they obey power and money. How right Ellul was about this deep complicity of business, politics, and the media that bamboozles populations!

So we collectively organized ourselves into an association called "Une Pointe pour tous" (one peninsula for all). But we also urged individual citizens to protest and rebel on their own, not just rely on the group or organization. The paradox is that an association can bring the impact of size and number but it also absorbs individuals and risks becoming a faceless entity. Stripped of its individuality and humanity it can become a mirror of the entity against which it fights.

Eventually, in January 2008, forty-two hundred of us signed a petition and sent it to our representatives and official authorities in the Médoc, Gironde, Aquitaine and France itself, all levels. We raised some money to make posters, signs, stickers, and tee-shirts bearing the message "NO to the Methane Terminal." We organized marches, rallies, and demonstrations. Unfortunately, these things didn't work all that well because it was winter and we were very few in number. But in the summertime, during holidays by the seaside, who wants to hear about worries, risks, industry or politics? Almost nobody! It's also very bad for shopping. We clearly confirmed what Jacques Ellul wrote in *De la Révolution aux révoltes* about the willing participation and submission of all citizens to the technician society in which they live.[2] Almost nobody cares. To tell the truth, we felt sometimes overwhelmed and helpless. Resignation and fatalism were recurrent when we talked with people in the street or on the beach.

Fortunately, we discovered Article 44 of the European Union Charter of Fundamental Rights: "Any citizen of the Union and any natural or legal person residing or having its registered office in a Member State has the right to petition the European Parliament." The key action was to have four individual citizen petitions sent to the European Petitions Commission. I prepared these four petitions with nearly identical texts denouncing the serious shortcomings of the Verdon project, notably in relation to several European "Directives" such as the "Birds Directive," "Water Directive," "Fauna and Flower Directive," and so on.

Then four of us went to Brussels on 31 March 2009 and the petition was officially declared "opened." Thanks to them our association was heard with all our supporting evidence. In a humorous way, we could say that the Pelobate toad saved us, much more efficiently than any of our representatives! Yes, in fact, this little animal living in the Verdon site is a protected species of frog. It is prohibited to kill or even displace one. For sure, this detail in our eventual victory helps to prevent us from being contaminated with our opponents' arrogance and to remain humble. We are precarious and fragile—alive and victorious because of a gift.

Another element that rallied the community against the project, ironically, was the NIMBY reflex felt by the Royan inhabitants on the opposite

2. Ellul, *De la révolution aux révoltes*, 89–90.

bank of the Gironde estuary! This was a much more populous area which became aware of, and then resistant to, changes in their landscape! According to one rumor, former French President Sarkozy's mother was one of those residents up in arms about it!

Here maybe is the point: chance and coincidence played a part in our story, alongside of our research, demonstrations, rallies, and petitions. It illustrates Ellul's caution and realism: there are very slim chances to win in a "democracy" ruled by technique and the technological system—slim, but not none, it's not non-existent. So we must hope and try. Hope is the point, whether you are Christian or not. Hope, Ellul said, is the bridge between the two parts of his work, sociology and theology.

We eventually forced the mainstream media to see us. We used European laws against French ones and local interests. But the odds were still stacked against us and even we can admit that normally we ought to have lost the fight. From the purely logical point of view, we would not have won, had we faced similarly engaged opponents. I like to think that we seemed so determined that we impressed them into thinking we could not fail! Something like a "democratic bluff" maybe. When I sent questions to the debate commission and obtained twelve or fourteen different answers for the same question, I demonstrated our opponents' lies. When I held up a placard with the Latin "CUI PRODEST?" (who profits by the crime?) in a meeting where I knew in advance that we wouldn't be allowed to speak, it demonstrated we were without illusion, but nevertheless without fear, not discouraged, and very present. (Actually, we were confused and in doubt, really close to desperate!).

In December 2007, at the last and biggest (900 persons!) public debate in Bordeaux, in December 2007, I insisted to our organizational leadership team that I speak. I had the unanticipated privilege of the last word on the program, when hundreds of people applauded with hope and some of them embraced me downstairs. Then I knew we had the very beginning of a chance to win.

We had a moral obligation to rebel because we had to protect people and nature from greed, destruction, and lies. Our deep convictions, our community together, and our formation by Jacques Ellul's ideas—all of that sustained us until our opponents eventually gave up and went away. They were not prepared like we were and they had wrongly believed they would have an easy time defeating us. Our Italian friends facing a similar project had told us that for them it took over ten years. Fortunately, there was a sort of miracle for us, it took only three years.

What gave us such a strength was the absolute certainty that we were fighting the good fight, we were "the good guys." Our righteous anger helped us and fueled our energy. Outraged, incensed by the contempt and arrogance of our opponent's propaganda, we knew we were defending justice and the future. We also knew we won a battle, not the war. Once the fight was over, we were left with the feeling that we couldn't just halt the destruction, we needed to create something positive. Therefore, we organized ourselves into another association, Agora Soulac Energie, to develop a true democratic debate and build positive solutions which could become local examples with very global impact. And we go on fighting against old school politics.

Franz Kafka wrote that a book—a good book—must be like a punch that jolts you out of the ranks of the murderers. Jacques Ellul's books have this inspiring power that changes your mind and thus changes your acts and your life. You are suddenly jolted out of your ordinary routine in thought as well as in action, and you begin considering what was usual and seemed necessary—as astonishing, unnecessary and questionable. You begin to accomplish that "ideological detoxification," that "demythization of the political for better thinking and acting" at which Jacques Ellul aimed.[3]

Bibliography

Compagnon, Daniel. "Foreword" to *L'Illusion politique*, by Jacques Ellul. 3rd ed. Paris: La Table Ronde, 2004.
Ellul, Jacques. *De la révolution aux révoltes*. Paris: La Table Ronde, 2011.
———. *Le Bluff Technologique*. Paris: Hachette, 1988.

3. Compagnon, "Foreword," 25.

11

Political Upheavals Old and New
The Radical Vision of Jacques Ellul

BY Langdon Winner

> *Langdon Winner (PhD, California) is Thomas Phelan Professor of Humanities and Social Sciences at Rensselaer Polytechnic Institute (USA). He is a political theorist interested in the ways technology affects the quality of social and political life. He is author of* Autonomous Technology, The Whale and the Reactor, *and other works, and his long-standing involvement with popular culture includes an earlier life as rock critic for* Rolling Stone. *His blog Technopolis offers commentary on political artifacts in humanity's uncertain future. At present he is at work on a book,* The Revolt Against Technology: A Study of Post–World War II American Social Thought.

* * * *

DURING THE PERIOD OF the 1960s in which Jacques Ellul's major writings were first translated into English and published for American audiences, the United States was embroiled in an ongoing series of social and political eruptions that might have offered a fertile field for the reception of the French thinker's ideas. The Civil Rights movement, environmental activism, anti-Vietnam War mobilizations, Free Speech Movement at U.C. Berkeley, subsequent student revolts on other college campuses, consumer rights awakening, Black Power and Black Panther insurgency, psychedelic rock and roll, Woodstock, the "counterculture," along with the women's movement, movement of people with disabilities, early signs of demands for LGBTQ rights—these along with numerous other expressions of cultural unrest could well have found sustenance in Ellul's forceful criticisms of the institutions, practices and obsessions of modern life. Indeed, the first of Ellul's works to appear prominently in American bookstores—*The Technological Society, Propaganda,* and *The Political Illusion*—did attract a good many readers among of those active in the quest for fundamental social change. But as compared to writings favored by activists of the time—Eric Fromm's edition of Marx's early writings, Paul Goodman's *Growing Up Absurd*, C. Wright Mills's *The Power Elite*, the "Port Huron Statement" of Students for A Democratic Society, Herbert Marcuse's *One-Dimensional Man*, Betty Friedan's *The Feminine Mystique*, and others—Ellul's books gained only a modest following.

There are many conceivable reasons why Ellul's approach did not catch fire with a much wider audience. It is possible that for activists of the time, the angle and character of Ellul's critiques did not seem directly relevant in attempts to confront the most obvious targets of concern: the hollowness of the consumer culture, growing dominance of the military-industrial complex, persistence of racial injustice, bureaucratization of higher education, sexual repression in middle class family life, and looming threat of nuclear annihilation. None of these were explicitly addressed in Ellul's writing, at least not in ways that most New Leftists, peace demonstrators, hippies, civil rights advocates, feminists, and environmentalists, and others would have found helpful in their search for meaning and quest for practical strategies of action. Indeed, the thrust of Ellul's inquiries on such matters and technology and social life, modern media and communications, the promise of modern revolutions, and the prospects for political reform tend to cast doubt upon most of the analyses and, especially, the remedies that even the most thoughtful activists of the period were inclined to embrace. In that sense, his radicalism was far too radical for many intellectuals and politicos to accept, precisely because it cast doubt on many of their preferred convictions and agendas for possible change.

Saying this, I do not mean to fault those who sought to challenge the underlying problems in American society or in the modern industrial order. But the insistent force of Ellul's observations and arguments simply cannot be packaged as a program, a social movement or road map for building a better society. A plausible reason why his writings did not have instantaneous appeal to the New Left, ant-Vietnam War militants, and the "counterculture" is that his views on both the manifestations of the troubles of modern society and of commonly proposed remedies for them make it clear that much deeper explanations and alternatives are urgently needed. Many of his arguments strongly suggest that the pathways of thought and action strongly favored by young radicals and prominent intellectuals were not only bound to fail, but actually mirror and tend to strengthen the very evils they were railing against.

For example, a common feature of student uprisings on college campuses of the time was to seek validation on television for the demonstrations and struggles of the day. "The whole world is watching" was a common chant and, in fact, demonstrators would often leave battles against campus administrators, the police and tear gas only to return to their dorm rooms and apartments to watch themselves on TV. "Hey, there we are on the screen!" was the rallying cry, taken to be a sure indication that one's role was highly significant. In contrast, Ellul's writings in *Propaganda* and other works of the period suggested that action primarily aimed at realization in electronic communication systems is futile at its core because it merely reinforces patterns of state power, a lesson perhaps even more painfully evident today than in the 1960s.

Although it was likely less studied and less widely appreciated than its two companion volumes, *The Political Illusion* was perhaps Ellul's most poignant statement about the underlying difficulties and prospects that confront citizens in modern liberal democracies. In the manner characteristic of many of his writings on society, economics, culture, and politics, much of the book is an extended commentary on the ideas and empirical findings of twentieth century social science in Europe and America. Perhaps following the maxim of Max Weber—who, when asked why he continued doing sociology, responded "I want to see how much I can stand"—Ellul dutifully plows through stacks and stacks of the mainstream literature in political science, public administration, communications research, and adjacent fields, seeking the key findings and gems of wisdom the research contains. Without distorting the conceptual or empirical foundations of the works he explores, Ellul argues that what the classics of twentieth century social science reveal are varieties of oppression and disconnection from reality that emerge within the basic, everyday institutions and practices of political society.

In much the same way that Marx claimed to have turned "Hegel on his head," Ellul takes the corpus of mid-twentieth century social science and turns it on its head, revealing not the realm of enlightenment and progress that post-World War II academic writers sought to describe, but a kind of twilight zone in which benighted citizens, politicians and bureaucrats wander helplessly in search of happiness, prosperity and security through involvement in the institutions of public life, institutions that turn out to be hollow of genuine, felicitous substance.

In the political science of the time, as Ellul notes, a key quest was to shed light on underlying institutional patterns of modern democracy, ones rooted squarely in economic, social and cultural interest groups engaged in the push and pull of electoral outcomes and intricate negotiations of policy shaping. Distinctly American narratives of this kind, in the Yale School of political science, for example, upheld the various structures, processes and dynamics of late twentieth century political life as the maturation of democracy within the supportive environs of electronic media, social psychology, public relations, methods of polling, public administration, and the like. The overall message was that such developments were tangible evidence of progress in institution building, accomplishments immune to the terrible excesses of "mass society" and totalitarianism that had ravished the middle century. Within the lively interactions of key groups and voting blocks in political pluralism, conflict took place in ways that ultimately resulted in peaceful accommodation achieved through graduated incrementalism, step by step development toward the good society. Fortunately, it was commonly held, the arrival of radio and television would enrich the process, cultivating a vast populace much better informed about elections, key issues and policy making than in previous generations. Among its adherents, this model was upheld as modern democracy at its very best.

Among many graduate students in the political science and sociology of the 1960s, however, the established disciplinary frameworks seemed increasingly unsatisfactory. Many were inclined to ask: Don't the pungent criticisms outlined by Students for A Democratic Society and thinkers of the New Left offer a better analysis of the real conditions in American politics? Why should we embrace theories that depict lifeless, abstract structures while ignoring the active presence of citizens? After all, where's the democracy? Indeed, the cherished prize seemed to have evaporated within all the number crunching and conceptual model building.

Although seldom part of the mainstream debate, Ellul's arguments addressed similar concerns. His chapter on "Participation" in *The Political Illusion* takes note of the ways that leading social scientists were advancing principles far removed from anything faintly resembling genuine democracy. Their basic contention, he noted, is that political society is composed of large well managed groups that interact to determine the shape of crucial public policies. "Finally," he notes, "in the general contemporary effort to return to democracy in some of its sense and value, there is an important trend among political scientists, ever since [Joseph] Schumpeter, which considers democracy not a choice among programs, ideas, or objectives, but only one among ruling groups, men charged with leading the game."[1]

---

1. Ellul, *Political Illusion*, 172.

In this light, one writer Ellul finds especially interesting is Seymour Martin Lipset, sociologist and author of the widely celebrated treatise *Political Man*. Ellul summarizes the book and its position succinctly.

> There still remains Seymour Martin Lipset's theory; a group of associations of oligarchic character contributes to maintaining democracy. For a society to be democratic, it is not necessary that the democratic rule be applied inside the organisms that constitute it. Unions, for example, represent the general interest of their members, who do better by joining unions than by remaining at the mercy of industry. . . . All the associations combined represent the divergent interests of all society; whereas every one of these associations limits the individual's freedom, it gives the leaders a much greater *real* freedom.[2]

At that point Ellul offers a wry comment. "This conception of democracy is really very touching, for it literally reproduces the description of feudal society."[3]

Through an ongoing stream of commentaries of this sort, Ellul dismantles the central ideas and arguments of academic social science of the period. To use a legal metaphor, his careful rendering of the social scientists' own brief for the defense of the virtues contemporary democracy generates compelling evidence for the prosecution. Again and again, Ellul argues that what are ostensibly open, democratic institutions are actually driven by motives of self-preservation, eventually achieving a status of "political autonomy" that makes them completely unrepresentative and unresponsive to the needs of the populace. His concept of "autonomy" points to the fact that key institutions of decision-making and administration have evolved to become tightly closed, immune to any effective outside influence.

The notion of "autonomy" in this setting evokes similarly provocative uses of the concept in *The Technological Society*. Of course, there are significant differences as well. As regards the autonomy of "la technique," Ellul took note of the ways in which a manic quest for efficiency, "the one best way" for accomplishing any objective, has infused all domains of human thought and practice in the modern age. Long a presence in earlier cultures, but limited by constraints imposed by religion, crafts guilds, political regimes, and other institutions, "la technique" was finally set free during the revolutions of modernity, emerging as viral presence that shapes forms of life from the largest organizational structures to the most minuscule features of matter and code. To use the language of today's digital systems, what Ellul identifies

2. Ellul, *Political Illusion*, 181.
3. Ellul, *Political Illusion*, 181.

as autonomous technique looms as a kind of universal "malware" that has infected and now commands just about everything that humans and their instrumentalities typically have accomplished in modern societies.

While by no means as universal or forceful a concept as that offered in *The Technological Society*, the notion of "autonomy" in *The Political Illusion* becomes the center of Ellul's insistent claim that political parties, elections, state institutions, civic organizations, political media, and all conventional varieties of citizen activity contain little that resembles democracy in a true sense. They have achieved "autonomy" and pursue their own goals far beyond anything a truly participative society would seek to achieve. The end result is to render what seem to be the primary vessels of political vitality completely empty of any genuine substance.

Although his judgments seemed extreme and pessimistic when first published, Ellul's reasoning actually foreshadows the findings of some of the most rigorous, systematic research in political science of the early twenty-first century. Deploying state of the art quantitative methodology, Martin Gilens and Benjamin I. Page have shown that the preferences of middle and low income people in the U.S. have little or no influence upon actual policy making. Comparing the desires of the populace as revealed by opinion polls to tangible results of the legislative process, they report "the preferences of the average American appear to have only a minuscule, near zero, statistically non-significant impact upon public policy." "In the United States, our findings indicate, the majority does not rule—at least not in the causal sense of actually determining policy outcomes." Careful not to overstate the blatantly obvious implications of their study, Gilens and Page conclude that "if policymaking is dominated by powerful business organizations and a small number of affluent Americans, then American's claims to being a democratic society are seriously threatened."[4]

Additional confirmation of Ellul's misgivings are evident in a number of recent political eruptions in the U.S. and Europe where grievances about pervasive oligarchy have become a common rallying cry. The Occupy Wall Street movement of 2011 and 2012 as well as the movement of "indignatos" in Spain that preceded it were energetic uprisings against political and business elites. By the same token, the 2016 Brexit vote to withdraw United Kingdom from the European Union signaled a widespread judgment that distant, unresponsive, self-interested bureaucrats in Brussels had lost touch with the needs and desires of everyday people and served only the priorities of society's richest strata. In somewhat similar ways the successful campaign of Donald Trump to become President of

---

4. Gilens and Page, "Testing Theories," 575–77.

the USA in the 2016 election expressed widespread discontent with the status quo, sentiments laced with the candidate's signature narcissism, racism, xenophobia and "reality" TV demagoguery.

As one ponders recurring disturbances of this kind, Ellul's judgments on the flaws in modern liberal democracy seem not only plausible, but increasingly prophetic. A key question looking forward is whether citizens will continue to embrace beguiling claims that economic growth as fueled by continuing technological innovation truly offers a better way of life? Or will burgeoning hordes of disgruntled souls revolt in realization that such promises are hollow—mere political illusions propagated by cloistered, self-interested elites in Wall Street, Washington, Silicon Valley, and Davos? As they survey the economic, technological, ideological, and political landscape that confronts them these days, a great number of people are ready to say: "Frankly, we're not buying it any more." For better or worse, much of the energy in recent "populist" eruptions, ones that arise on both the "left" and "right" ends of the political spectrum, confront today's politicians, businessmen, and the mass with challenges to their authority, insurgencies that must somehow be deflected, redirected, tamed, or otherwise constrained.

In an especially revealing passage in *The Political Illusion* Ellul argues that the theory of economic alienation in Marxism along with remedies of economic democracy proposed on the left no longer address a much deeper predicament that faces humanity. The social, political, technological, informational order that surrounds us now claims our very souls and neutralizes people's best inclinations, their ability to think and act in meaningful ways. He writes, "Now the problem is for the powers that be—political or technological—to possess man internally, to organize fake appearances of liberty resting on fundamental alienation, to fabricate false appearances of personality resting on integration and radical massification."[5] In certain respects this critique points towards a general response characteristic of the "counter-culture" of the 1960s and 1970s, an amorphous social movement that, despite its notorious flaws and fantasies, was predicated on the rejection of middle class conformism, mindless consumerism, conventional varieties of employment, and a concerted search for alternative, more satisfying ways of living. The ultimate demise of this colorful upheaval bears witness to the range of insidious power to which Ellul's vision of modernity bears witness. Merely changing one's clothing, moving to a commune, listening to psychedelic rock, and using "controlled substances" were not sufficient to alter the deeply rooted patterns of the "American Dream" that claimed one's being.

---

5. Ellul, *Political Illusion*, 233.

Anyone who read *The Political Illusion* during that period would have seen the reasons for such dead ends spelled out in intricate detail.

While Ellul does not go so far as to advocate explicitly a mass revolt against ways of living structured upon materialism, efficiency, propaganda, cultural uniformity, and ingeniously enforced compliance, his book suggests that more serious uprisings would be fully justified. His brief comments throughout the book, especially the chapter on "Man and Democracy" indicate what an appropriate response would be.

It amounts to a suggestion that each person step outside the stagnant oppressive economic, political and technological milieu that surrounds us all. Any credible movement in that direction would begin the identification and open discussion of what he calls "tensions" in society, ones that divide people one from another and yet offer opportunities for dialog, mutual respect and action. Examples of such tensions from earlier periods of history certainly include the tension between church and state as well as between the bourgeoisie and laboring people, tensions that have inspired a variety of attempts to democratize society in modern history. Ellul implies that people today would have to identify significance tensions that have arisen in the present time, points of "differentiation" and possible contention in their lives and surroundings. Such steps would certainly include concerted efforts to rescue the powers of language and reason from the clouds of toxic fog produced each day by the communications systems of the modern technological state. He insists that the "reasonable man, without whom human democracy cannot exist, is the one who at the same time can restore to language its true reason and its communicative substance. . . . The more that language loses its content and reasonable structure, the more man is delivered to propaganda's delirium."[6]

While the implications of this advice seem perfectly clear as regards possible rebellions by individual women and men, what does Ellul recommend as steps toward discovering, clarifying and ultimately affirming matters that we—as humans and as members of geographically situated societies—have in common? To recognize a life in politics as something other than contending with a long list of oppressive conditions and collective maladies would certainly require identifying more promising pathways. Indeed, as the book moves toward its conclusion, Ellul argues that any genuinely democratic politics involves the classic question: How are we to live together?

He writes, "The common measures of what we have to say to one another and of what makes communication possible, of what we jointly have

---

6. Ellul, *Political Illusion*, 236.

to live for . . . must be constantly rediscovered and recreated."[7] "We must understand that democracy is always infinitely precarious and mortally endangered by every new progress. It must be forever started again, rethought, reconstructed, begun again."[8]

What he offers, then, is a very stern challenge, one that sets a very high bar for the attainment of anything remotely resembling a democratic way of life. Rather than seek to devise programs, initiatives, platforms, and new organizations with new agendas, he insists that true citizenship must involve continual, persistent discussion, debate, and critical thinking to clarify who "we" are and what "we" are and what "we" hope to do in the first place. Then and only then might "we," as truly free citizens, carefully proceed to build institutions, make laws and formalize arrangements for a just, reasonable political society.

Indeed Ellul is not especially optimistic that his generation or any later ones would realize the conditions offered in his sketch. The traps and pitfalls are many and people are drawn to them like flies to flypaper. Unlike the plans often rolled out in the glib musings of modern social science, he depicts democracy as something very difficult to obtain, something often advertised but seldom realized, something extremely fragile and subject to cynical abuse.

"If man were left to himself," he writes, "his inclinations, his responsibilities, his personal choices, on his own level, without systematic influence, propaganda, 'human relations,' group dynamics, obligatory information, directed leisure, then slowly, humbly, modestly, democracy might perhaps be born."

Then he adds, "But how newborn, how weak and fragile it would be!"[9]

Bibliography

Ellul, Jacques. *The Political Illusion*. Translated by Konrad Kellen. New York: Knopf, 1967.

Gilens, Martin, and Benjamin I. Page. "Testing Theories of American Politics: Elites, Interest Groups, and Average Citizens." *Perspectives on Politics* 12 (Sept. 2014) 575–77.

---

7. Ellul, *Political Illusion*, 237–38.
8. Ellul, *Political Illusion*, 230.
9. Ellul, *Political Illusion*, 232.

12

Christian Political Engagement in a New Key?
Reading Ellul in Ottawa[1]

BY Paul C. Heidebrecht

> *Paul C. Heidebrecht is Director of the Centre for Peace Advancement and Adjunct Professor at Conrad Grebel College, University of Waterloo (Ontario, Canada). He previously was the Ottawa Office Director of the Mennonite Central Committee of Canada. He is a graduate of the University of Waterloo (BAS, Mechanical Engineering), the Anabaptist Mennonite Biblical Seminary (MA, Theology & Ethics), and Marquette University (PhD, Religious Studies). With the Mennonite Central Committee, Paul worked as an Appropriate Technology Engineer in Bangladesh and on peacemaking initiatives in Nigeria.*

* * * *

I FIRST ENCOUNTERED THE work of Jacques Ellul in the halls of academia, long before I moved to Ottawa to direct the Mennonite Central Committee (MCC) office in Canada's capital city.[2] As an engineering student, and then in the course of researching seminary and graduate school theses in Christian ethics that focused on the topic of technology, I was exposed to both his social scientific and theological writings.

---

1. An earlier version of this essay was presented at "Prophet in the Technological Wilderness: A Centenary Celebration," International Jacques Ellul Society Conference at Wheaton College, Illinois, July 10, 2012.

2. MCC pursues "relief, development, and peace in the name of Christ" on behalf of North American Anabaptist churches. Established in 1920, MCC currently works in over fifty countries around the world, including Canada and the USA.

Thus I came to the work of political engagement with a deep appreciation for Ellul's analysis of technique, and his appropriation of the biblical concept of the principalities and powers. For example, I was convinced that the church was called primarily to bear witness to Christ's defeat of these powers, rather than seek their redemption. And I was wary of the impulse to grab hold of the "handles of history" to make things come out right, an impulse embodied in many of the social justice activists from mainline, evangelical, and Roman Catholic churches that I found myself partnering with in Ottawa.[3]

My wariness has not departed. It is clearer than ever to me that, as Ellul insisted, Christian faith and revelation are what makes it possible to overcome paralysis in the face of a close analysis of our current sociopolitical context. Having said this, it might strike you as somewhat incongruous that someone with even a hint of Ellul's sensibilities would be inclined to encourage engagement with the government of Canada, rather than simply ignoring it.[4] This chapter is an attempt to explain why it might make sense for churches and church agencies in the Canadian context to pay close attention to governmental structures. My thesis is that the pursuit of political engagement is one important way for Christians to act in the midst of a barren moral landscape defined by technique.

In what follows I will briefly define what I mean by political engagement, before describing unique dimensions of the Canadian political context. I will conclude by articulating a vision for the work of political engagement that grows out of my experience with MCC. In the process I intend to provide signals as to how this vision has been informed by, and goes beyond, my reading of Ellul.

---

3. Another way to put this is that I was suspicious of the pattern of behavior that, according to American social theorist James Davison Hunter, characterizes both the Christian Right and Christian Left: they presume to "know God's specific plans in human history and that [we] possess the power to realize those plans in human affairs." See Hunter, *To Change the World*, 95. I think the vision articulated in this chapter bears a strong resemblance to Hunter's proposal that Christians focus on being a "faithful presence."

4. An overall thrust to Ellul's work is that the private life should take priority over public or political action. After all, as he once put it, "There is no legitimate political authority as such. Political authority and organization are necessities of *social life* but *nothing more than necessities*. They are constantly tempted to take the place of God." See Ellul, "Anarchism and Christianity," 22. For an expanded discussion of this point, see Ellul, *Anarchy and Christianity*.

Defining Political Engagement

Within churches, church agencies such as MCC, and other non-governmental or civil society organizations, the expression *political engagement* is often used synonymously with the term *advocacy*, although it also includes government relations, public affairs, or lobbying. Political engagement is now commonly viewed as one of several tools that can be utilized by organizations in order to achieve larger objectives, alongside things such as material resources, and efforts to build the capacity of partners. There may be times of crisis when needs can only be met with emergency relief. And there may be times when a longer-term approach is utilized in order to achieve the development objectives of a community. But there are also times when the root causes of a crisis or the barriers to development transcend a local context. In the words of Christian activist Shane Claiborne: "We give people fish. We teach them to fish. [But we also] tear down the walls that have been built around the fish pond. And we figure out who polluted it."[5]

MCC frames its advocacy or political engagement work as "a form of public witness and a tangible way of loving our neighbor."[6] Staff are compelled to speak out because of relationships that have been established with diverse partner organizations and communities in Canada and around the world—partners and communities who lack the same kind of access to governmental power structures. Thus the priorities and policy positions pursued emerge out of practical, on-the-ground experience, rather than theoretical analysis alone. In the past it has been argued that it is this unique vantage-point—not the size or political views of their constituency—that has given MCC authority and credibility with policy-makers.[7]

Describing the Canadian Political Context

There are numerous (and some quite well-worn) distinctions that can be drawn between Canada and the United States. Obviously, a constitutional monarchy with a Westminster-style parliamentary democracy functions differently than a constitutional republic with separate legislative and executive branches of government. For starters, power is concentrated in fewer hands. And certainly Canada, as the smallest member of the G-7, plays a

---

5. Claiborne, *Irresistible Revolution*, 150–51.
6. MCC Advocacy Offices, "Loving Our Neighbor."
7. This has been confirmed directly by senior policy-makers over the years, not to mention politicians as prominent as former Prime Minister Pierre Trudeau—see Kreider and Goossen, "Listening and Speaking," 322.

very different role in the world than a global superpower does. Power may be more concentrated, but there is much less of it! Most importantly, for the purposes of this chapter, despite having a close trading relationship and ready access to American news and entertainment, there is a strikingly different attitude toward government in Canada.

One signal of this can be seen in the differing myths that aim to unite citizens north and south of the 49th parallel. While Americans conceive of a nation defined by the phrase *life, liberty, and the pursuit of happiness*, Canadians talk about *peace, order, and good government*. In fact, as philosopher and public intellectual John Ralston Saul points out, official references to this Canadian expression in the first decades of the twentieth century spoke of peace, *welfare*, and good government.[8] In general, Canadians have a more positive view of government, and many would find the suggestion, now commonplace among many American theologians, that "the nation-state is simply not in the common good business" rather odd.[9] After all, Canadians take considerable pride in the fact that their federal government funds public goods such as health care for all its citizens.

Governments in Canada not only pay for more—proportionally, of course—health care workers and public broadcasters, but more teachers, more social workers, and more regulators than governments in the USA do. While this is offset by the fact that governments in Canada pay for fewer military personnel, police officers, and prison guards, the overall level of public sector employment—and taxation—is still significantly higher in Canada than in the USA. Government bureaucracy is proportionately bigger, more complex, and more intertwined with the lives of more people. One might reasonably expect, therefore, that in Canada there would also be a proportionately greater "preoccupation with technocracy," as Ellul once warned. That it would be more likely to find "that the politician would yield to the technician."[10] One might think that the specter of what Foucault termed "governmentality" would loom large.[11] But this has *not* been the most striking problem that I have observed. I will say more about this in a moment, but the bigger problem in recent years at the federal level of government in Canada seems to be the extent to which the technicians have been yielding to the politicians. Ideology appears to be trumping technocracy.

---

8. Saul, *Fair Country*, 114–15.

9. Cavanaugh, *Migrations of the Holy*, 42.

10. Ellul, "Technique," 39. Not only the scale but the degree of centralization of political power was the focus of much of Ellul's concern. See, for example, Ellul, "Decentralism," 14.

11. See Dean, *Governmentality*.

Churches have also functioned in a different way in the Canadian political context than they have in the USA. When MCC established a presence in Washington, DC, in 1968, it was following a path already well-worn by numerous other churches and church agencies. In contrast, when MCC established a presence in Ottawa in 1975, it was a pioneering move among Canadian churches and church agencies, who were accustomed to relating to the federal government from their head offices in cities like Montreal and Toronto.[12] This is not to say that Canadian churches never had the ear of those in power. At the time of Confederation in 1867, Protestant church leaders exerted considerable influence over Anglophone politicians, and Roman Catholic bishops exerted considerable influence over Francophone politicians. The crumbling of these ties happened gradually over the course of the twentieth century in English-speaking Canada, along with growing ethnic and religious diversity, and with lightning speed in Quebec through the "Quiet Revolution" of the 1960s.

The resulting embrace of secularism by political elites and the media meant that churches found themselves on the outside looking in by the 1970s, and thus many other denominations, ecumenical organizations, and religious lobbyist groups have ended up joining MCC in having a presence in Ottawa since then.[13] My larger point in recounting this history is that both the reality and the demise of Christendom—the binding and subsequent loosing of the connection between religious and political power—is clear in the Canadian context in a way that it may not be in the USA.[14]

This post-Christendom era has been nuanced in recent years by a post-secular shift that has created a new kind of space for religion in the public square in Canada.[15] It is increasingly recognized that Canada is best

12. Kreider and Goossen, "Listening and Speaking," 328.

13. Several inter-church coalitions also formed during this period in order to pursue advocacy around a number of social justice issues, and MCC represented Mennonites in several of them. See Lind and Mihevc, *Coalitions for Justice*.

14. I am presuming that this shift is a good thing, for reasons articulated by, for example, Doerksen in *Beyond Suspicion*. Nonetheless, I recognize that this is still something that is open for debate in part because of the work of theologians such as John Milbank—see "Power is Necessary."

15. I am using the expression post-secular in the same way that philosopher James K. A. Smith does, meaning that religion now has a place in the public square in a way that modern theories of secularization cannot account for. See, for example, *Introducing Radical Orthodoxy*.

Like Smith, I am persuaded by Charles Taylor's thick account of the secular age, which distinguishes three facets of secularity: (1) the retreat of religion in public life; (2) the decline in belief and practice; and (3) the change in the conditions of belief. Taylor's description of what he terms "secularity 3" is an apt description of the post-secular: "Religious belief now exists in a field of choices which include various forms of

described as a mosaic not only of different linguistic groups, cultures, and ethnicities, but also of different religious worldviews. Religion is no longer viewed as something Canadians are expected to leave at the door when they enter into public discussions. And the modern liberal assumption that a harmonious society requires a shared, universal worldview seems to have been jettisoned in favor of a relatively secure pluralism in which particular beliefs are both legitimately irreconcilable *and* legitimately contestable.

Of course, this shift has also brought with it challenges. For example, the presence of Christian groups asserting themselves in the public square over hot-button social issues has led some to worry that a USA-style "religious right" movement is emerging in Canada,[16] or, at the very least, that political polarization is being exacerbated by (and replicated among those involved in) Christian political engagement.[17] Another challenge has been the fact that governments are increasingly suspicious of religious organizations of any kind that claim to speak on behalf of a group of individuals. The ability of political parties to slice and dice the population based on their own research and analysis means that they likely have a better idea of what people in the pews really think about a given issue than denominational leaders do. Thus it seems that the voice of churches has been diminished, even as the religious worldviews of individual Canadians are being taken into greater consideration.

It is clear that the growing sophistication—and effectiveness—of new election campaign strategies and tactics has meant they are now being applied to governing as well as campaigning. Governments are less concerned with policies that represent the views of their citizenry as a whole than they are with maintaining support from their base, and building support among carefully targeted swing voters. Thus the government's communication strategy often seems to be aimed at building wedges within the population rather than homogenizing it. Although Ellul's claim that "the aim of modern propaganda is no longer to modify ideas, but to provoke action"[18] still holds true, I am not so sure whether it "must capture the individual and the mass

---

demurral and rejection," which highlights the contestability of belief, but also the potential "for recompositions of spiritual life in new forms." See Taylor, *Secular Age*, 437.

16. See, for example, McDonald, *Armageddon Factor*.

17. See Blaikie, *Blaikie Report*; and Gruending, *Pulpit and Politics*.

18. Ellul, *Propaganda*, 25. Ellul makes a similar point in "Obstacles to Communication": "The purpose of language in propaganda is no longer to communicate ideas and rational thoughts but to provoke reactions and reflexes" (406). Indeed, the "absolute devaluation of language" has been evident in recent Canadian political discourse through the simplistic application of trigger words in order to create a climate of fear and distrust.

at the same time."[19] I am also left wondering about Ellul's insistence that "propaganda no longer obeys an ideology."[20]

Furthermore, in Canada the centralization of executive and legislative power in the cabinet, and, in particular, the office of the Prime Minister, means that partisan political appointees and party officials are able to exclude voices even from within the government's own bureaucracy from decision-making processes. Time and time again in recent years government ministers have made it clear they have no interest in being informed by statistical or scientific data that doesn't serve their political needs during policy debates—prominent examples include a tough-on-crime approach to the criminal justice system, enthusiasm for the expansion of extractive industries, and significant military procurement decisions.[21] In short, Canada has witnessed the triumph of technique that is in the service of ideological interests.

Political Engagement Reexamined

At this point it may be rather obvious that political engagement is fraught with risks. Indeed, warning bells may have been going off from the outset when I defined political engagement in rather instrumental terms, referring to it as a "tool" and positioning it as a means of achieving the greater end of program objectives. Rather than being tempted to view themselves as competing on the same plane with the principalities and powers, you may be thinking that churches need to stress their distinctiveness. Rather than comparing communications strategies and tactics—and sharing a preoccupation with finding the most effective and efficient approach—churches need to frame their work in a different way.[22]

I do not disagree with these reactions. And I would insist that my approach to political engagement is not motivated by any illusions of

---

19. Ellul, *Propaganda*, 91. Of course, it is still true that partisanship "negates the individual and reduces him to a cipher" (90).

20. Ellul, *Propaganda*, 196. This is evident in the lack of truthfulness in political advertisements—it doesn't always seem as though, as Ellul once put it, "truth pays" and that "propaganda must be based upon facts." See "Information and Propaganda," 62.

21. Of course, data is always value-laden and should never be presumed to be objective. My concern here is the extent to which the definition of political judgment has narrowed—as noted by Philip E. Tetlock, it has come to serve the needs of partisan spectators preoccupied with the drama of politics itself. See *Expert Political Judgment*.

22. Of course, given the difficulty that churches and church agencies have in resisting the lure of technique, should it be any surprise that they would be susceptible to this temptation in the realm of political engagement?

grandeur, but rather the opportunities that arise on the margins of Canadian political debates. As much as organizations such as MCC have focused on the big picture, on systemic or structural causes of injustice, the points of access to these systems and structures are not always at a very high level, and are not always obvious.

Sometimes, for example, these opportunities arise out of connections that MCC has established with a government department in the course of carrying out program work. Notable examples include receiving funding for relief, development, and peace work from Global Affairs Canada, negotiating agreements with Immigration, Refugees and Citizenship Canada in order to create mechanisms for refugee resettlement, and piloting restorative justice initiatives within Correctional Services Canada.[23] This work may be appropriately characterized as tinkering at the margins of systems that are fundamentally flawed. But there are clearly cases where this tinkering has led to concrete progress toward greater peace and justice—if not for the long term, then at least for an interim period.

Program connections with government are an example of the ad hoc nature of political engagement; it *does* require systemic thinking and analysis—being able to maneuver within and alongside existing institutions, and at times to even have a hand in creating new ones—but it does *not* presume that we possess systemic or comprehensive solutions. Like Ellul, I disavow system building. As he once said, "I have refused, you might say systematically, to produce a system."[24] Furthermore, these connections make the limitations of governmental systems and structures abundantly clear.[25] After all, sometimes it takes the effort of pursuing shared work, not simply theoretical conversations, to be able to get to the point where we realize precisely where it is that we differ. Another line from Ellul that continues to resonate comes from interviews originally broadcast on CBC Radio in 1981: "For me it's

---

23. Further discussion of these partnerships can be found in: Bietz, "MCC Canada Experience," and "Statement of Values"; and Janzen, "1979 MCC Canada Master." For Ellul, these kinds of connections would likely be troubling: "Becoming a partner to the State it has turned the State into another Church" (Ellul, "Anarchism and Christianity," 22). Thus, it is crucial to note that all three of these connections emerged organically as the result of grassroots initiatives that bumped up against the state; they were not strategically pursued by MCC.

24. Ellul, *In Season*, 203.

25. Governments are complex and dealing with their bureaucracies can be time consuming and frustrating. They also have tremendous inertia, and therefore are unable to react or change direction as quickly as the circumstances may warrant. Finally, the frame of reference that we are working within means that there are times MCC has come to the conclusion that the government's vision is fundamentally incompatible.

always the same problem. *Intellectual interest means concrete commitment, practical and political involvement.*"²⁶

A second kind of opportunity for political engagement at the margins arises in the legislative process. Laws, beyond the obvious realm of the criminal justice system, regulate businesses and professions, and provide a framework for the delivery of countless services. They define the parameters within which a society operates, and thus are the epitome of the ordering function of government. In Canada, the legislative process also provides the opportunity for Parliamentarians outside of the government's cabinet to address issues of concern to citizens. Known as Private Members' Bills, these initiatives rarely culminate in the passing of a new law, but they have had other kinds of impacts. In recent years there have been several examples of this kind of legislation that have been of particular interest to organizations such as MCC, including legislation seeking to create a peace tax fund for conscientious objectors to war, legislation to make it possible for U.S. war resisters to remain in Canada, and legislation to establish a Department of Peace.²⁷ A closer examination of the circumstances of all of these bills makes it clear that the pursuit of what some have called "pacifist-friendly legislation"—and others have dismissed as the attempt to "legislate peace"—is not what it appears to be at first glance. Many of these initiatives are not actually pursued with the expectation that laws will be changing anytime soon. And they are certainly not pursued with a naïve trust that changing laws will automatically result in social change.

At the same time, however, they are more than symbolic gestures. They are pursued because individuals and organizations have felt compelled—or in some cases have even been invited—to frame an ethical argument using the language of legislation. Put another way, they have provided an occasion to articulate deeply held moral convictions.²⁸ Not the only occasion, to be sure. And legislation is certainly not the first, or the most natural, language with which to speak. But I would argue that it is not an occasion to be dismissed, because the hard work of bearing witness to our faith in this way is, in itself, of tremendous value. It not only requires expertise, but careful consultation with those who possess on-the-ground experience. Most importantly, not

26. Ellul, *Perspectives on Our Age*, 25.

27. For an in-depth discussion of these bills, see: Heidebrecht, "Reframing."

28. In fact, legislation presumes the possibility of making moral arguments in the House of Commons. The issues that are debated by Parliamentarians are not confined to the realm of political, economic, legal, and scientific calculation, but also include moral considerations. And, however rare they may be, there are times when elected officials make decisions that are not focused solely on their own—or even their nation's—self-interest.

only does this effort make a contribution to public debates, it makes a contribution to internal debates within the churches that support this work. It provides an occasion to ask ourselves what being a follower of Christ in the Canadian context really means. What does it mean to be a conscientious objector to war? What does it mean to say we wish the government would pursue more creative approaches in working for peace and justice? In the words of Stephen Webb, "Christians do not know what they really believe until they publically witness to their faith."[29]

Indeed, my approach to political engagement has been motivated by more than the desire to achieve political change. More than reshaping the barren moral landscape of society, partnering with the Canadian government in program initiatives and participating in the legislative process has the potential to enliven the moral landscape of the churches that support organizations such as MCC. I think that this kind of political engagement should properly be viewed as creating space for Christians to embody the gospel of peace through the process of wrestling with the impact of actual, not theoretical, systems and structures. And so our attention is misplaced when we focus on effecting change in governments as the primary objective of political engagement. Put another way, I would argue that the best measure of Christian political engagement is *polity* influence, not policy influence. The most profound impact is ecclesial, not governmental or societal.[30]

To conclude, in my view there are times when Christians *should* pay close attention to governmental structures, and not simply for the purposes of analysis and critique. Even in the midst of principalities and powers defined by the triumph of technique, I think people of faith are called to put

---

29. Quoted by Hovey, *Bearing True Witness*, 186. Hovey notes that Webb "follows Barth, who describes how bearing witness to the content of faith is the way that 'cognizance' becomes 'knowledge.'"

30. Of course, we need to be careful that we don't talk about political engagement simply as an instrument, tool, or technique working toward a different kind of end. Following Craig Dykstra, I think the point of Christian practices, unlike most practices, is not human achievement or excellence:

"Christian practice is different. And that is because its story is different. While human achievement is valued in the Christian story, it has a different place and meaning. The human task is not fundamentally mastery. It is rather the right use of gifts graciously bestowed by a loving God for the sake of the good that God intends—and ultimately assures. . . . So our basic task is not mastery and control. It is instead trust and grateful receptivity. Our exemplars are not heroes; they are saints. Our epitome is not excellence; our honor is in faithfulness" (Dykstra, *Growing in the Life of Faith*, 76).

Like Dykstra, I am convinced that Christian practices do not assume their formative power in a straightforward cause-and-effect fashion, but precisely because it is the Holy Spirit that is working through them. However, insisting that making Christian disciples is the job of the Holy Spirit does not mean that we can say nothing about the process through which this happens.

flesh on an alternative ethical vision. We are called to concrete—albeit ad hoc—action. And the intent or focus of this action may be directed toward a different kind of change than what is typically envisioned.

Bibliography

Bietz, Ron. "MCC Canada Experience: A Case Study of Government Funding for Development," and "Statement of Values and Guiding Principles Governing MCC Canada's Relationship with CIDA and Use of CIDA Funds." *MCC Peace Office Newsletter* 25/4 (October–December 1995) 6–9.

Blaikie, Bill. *The Blaikie Report: An Insider's Look at Faith and Politics.* Toronto: United Church Publishing, 2011.

Cavanaugh, William T. *Migrations of the Holy: God, State, and the Political Meaning of the Church.* Grand Rapids, MI: Eerdmanns, 2011.

Claiborne, Shane. *Irresistible Revolution: Living as an Ordinary Radical.* Grand Rapids, MI: Zondervan, 2006.

Dean, Mitchell. *Governmentality: Power and Rule in Modern Society.* 2nd ed. London: Sage, 2010.

Doerksen, Paul. *Beyond Suspicion: Post-Christendom Protestant Political Theology in the Thought of John Howard Yoder and Oliver O'Donovan.* Eugene, OR: Wipf & Stock, 2010.

Dykstra, Craig. *Growing in the Life of Faith: Education and Christian Practices.* Louisville: Geneva, 1999.

Ellul, Jacques. "Anarchism and Christianity." *Katallagete* 7/3 (Fall 1980) 14–24.

———. *Anarchy and Christianity.* Translated by Geoffrey W. Bromily. Grand Rapids, MI: Eerdmans, 1988.

———. "Decentralism: Is It Possible as a Central Government Policy?" *Raise the Stakes: Planet Drum Review* 1/3 (Summer 1981) 14.

———. *In Season, Out of Season: An Introduction to the Thought of Jacques Ellul.* Based on interviews by Madeleine Garrigou-Lagrange. Translated by Lani K. Niles. San Francisco: Harper and Row, 1982.

———. "Information and Propaganda." *Diogenes* 5 (1957) 61–77.

———. "The Obstacles to Communication Arising from Propaganda Habits." *Student World* 52/4 (1959) 401–10.

———. *Perspectives on Our Age: Jacques Ellul Speaks on His Life and Work.* Edited by William H. Vanderburg, translated by Joachim Neugroschel. New York: Seabury, 1981.

———. *Propaganda: The Formation of Men's Attitudes.* Translated by Konrad Kellen and Jean Lerner. New York: Vintage, 1973.

———. "Technique, Institutions and Awareness." *American Behavioral Scientist* 11/6 (1968) 38–42.

Gruending, Dennis. *Pulpit and Politics: Competing Religious Ideologies in Canadian Public Life.* Calgary: Kingsley, 2011.

Heidebrecht, Paul C. "Reframing Mennonite Political Engagement as Christian Formation." *Conrad Grebel Review* 34/1 (Winter 2016) 74–95.

Hovey, Craig. *Bearing True Witness: Truthfulness in Christian Practice.* Grand Rapids, MI: Eerdmans, 2011.

Hunter, James Davison. *To Change the World: The Irony, Tragedy, and Possibility of Christianity in the Late Modern World*. New York: Oxford University Press, 2010.

Janzen, William. "The 1979 MCC Canada Master Agreement for the Sponsorship of Refugees in Historical Perspective." *Journal of Mennonite Studies* 24 (2006) 211–22.

Kreider, Robert S., and Rachel Waltner Goossen. "Listening and Speaking: The Washington and Ottawa Offices." In *Hungry, Thirsty, a Stranger: The MCC Experience*, by Robert S. Kreider and Rachel Waltner Goossen, chapter 25. Waterloo, ON: Herald Press, 1988.

Lind, Christopher, and Joseph Mihevc, eds. *Coalitions for Justice: The Story of Canada's Interchurch Coalitions*. Ottawa: Novalis, 1994.

McDonald, Marci. *Armageddon Factor: The Rise of Christian Nationalism in Canada*. Toronto: Random House Canada, 2010.

Mennonite Central Committee (MCC) Advocacy Offices. "Loving Our Neighbor through Witness to Government." Akron, PA: Mennonite Central Committee, 2009.

Milbank, John. "Power Is Necessary for Peace: In Defence of Constantine." *ABC Religion and Ethics blog* (29 October 2010).

Saul, John Ralston. *A Fair Country: Telling Truths About Canada*. Toronto: Penguin, 2008.

Smith, James K. A. *Introducing Radical Orthodoxy: Mapping a Post-Secular Theology*. Grand Rapids, MI: Baker Academic, 2004.

Taylor, Charles. *A Secular Age*. Cambridge, MA: Belknap, 2007.

Tetlock, Philip E. *Expert Political Judgment: How Good Is It? How Can We Know?* Princeton: Princeton University Press, 2006.

13

Bringing Ellul to the City Council
Mayor Robb Davis Reflects on Ellul's Influence

BY Mark D. Baker

> *Mark D. Baker has an MA in Biblical Studies from New College Berkeley and a PhD in Theology and Ethics from Duke University. He is professor of mission and theology at Fresno Pacific Biblical Seminary. He has written a number of books and articles in English and Spanish, including some that interact with Ellul's thought, such as*: Religious No More *(InterVarsity, 1999) and* Gálatas *(Kairos, 2014).*

* * * *

JACQUES ELLUL'S BOOKS, THE *Political Illusion* and *The Politics of God and the Politics of Man* were based not only on observation of the political process, but also on his experience as a participant. He served on the Bordeaux city council from August 1944 until April 1945.[1] How could one read these, and other books by Ellul, and then decide to seek the very office that contributed to Ellul's critical view of politicians and political activity? I asked that exact question when Robb Davis, an avid reader of Ellul, told me he was running for city council.

Robb Davis holds a master's degree in public health and a PhD in population dynamics from Johns Hopkins University. He has over twenty years experience in international development in the field of maternal and child health and nutrition. He was the executive director of the Mennonite Central Committee. He contributed an article to the *Ellul Forum* (#46). He was

---

1. For more information about and comments by Ellul on his experience in a city administration see: Ellul and Garrigou-Lagrange, *In Season*, 52–56; Ellul and Troude-Chastenet, *Jacques Ellul on Politics*, 79–85.

elected to the Davis, California, city council in June 2014 and began serving as mayor of Davis in July 2016. I had the opportunity to interview Robb during a plenary session of the 2016 conference of the International Jacques Ellul Society. Part one of this chapter is an edited version of excerpts of that session. [2] Part two is a follow-up interview I did on July 3, 2017.

The Interview: Part One

Mark: It would be surprising to many that an enthusiastic reader of Jacques Ellul would run for political office. How did Ellul's work factor into your decision to run for city council?

Robb: I'll start by that saying Ellul arguably is the reason I became involved in city politics. Maybe even more surprising than my claiming to have run for office on the basis of something Ellul said, which many might consider to be paradoxical, is that I am also a Mennonite. I wasn't just trying to break some molds. I had spent about twenty-five years travelling the world. I was a technician, dispensing wisdom to many villages and communities all over the planet, forty-five different countries. I started reading Ellul and Patrick Deneen and they challenged me about living and acting locally. So about seven years ago, I stopped travelling. I decided not to get in an airplane anymore. And that changed everything.

When I started digging into my hometown I realized that the brokenness that I had experienced other places was actually more profound in Davis, California. We had a veneer of privilege and beauty, but not too far below the surface we had serious problems of addiction, homelessness, racism, and exclusion. While staffing an overnight shelter, I saw how we fail as a society to treat mental health and addiction, and how these things leave people broken. I saw how we toss the problems over to the nonprofits to try to figure out a solution.

Where I drew from Ellul quite a bit, was the idea of the flourishing of intermediating entities outside the state. The state was incapable, even at a local level, of effectively dealing with these problems. Into the interstices came these small organizations. My commitment at that time was to try to work with them to make them stronger, to try to take some things I'd learned in my trips around the world and bring them into the community. When you do that for a while people will often ask you to be on a commission or a task force. Then somebody knocks on your door one day and

---

2. Part One was published in *The Ellul Forum* 58 (Fall 2016) 1–8.

says, "Maybe it would be useful for you to run for office." I didn't believe that I should or could do it.

My main concerns related to the issues Ellul raises about power. The thing that pushed me towards the decision was the idea that perhaps in that role I could encourage the flourishing of these intermediating agencies in the community. This gets back to power. One reality of being a political leader is, when you pick up the phone and say to someone, "Come to a meeting," they'll come. I thought, "Maybe I can bring people around the table who aren't talking to each other, maybe I can bring the school district together with the police department, together with the city, and do a restorative justice program."

Another key factor that led me to run was born out of something I read in Ellul: "A key fact of this civilization is that more and more, sin has become collective and that the individual is constrained to participate in it."[3] I thought that if we had someone in office who was engaging in regular confession about our participation in that collective sin, maybe that would be helpful to a community.

Mark: How did Ellul influence your campaign, how you ran for office?

Robb: In *The Technological Society* Ellul, commenting about propaganda, states: "Whether technique acts to the advantage of the dictator or the democracy it makes use of the same weapons, acts on the individual, manipulates his subconscious in identical ways, and in the end leads to the formation of exactly the same type of human being."[4] I saw that people running for office, even locally, used propaganda for very specific ends, which is the building of allegiance toward themselves. I was told, "You've got to sell yourself. This is about you, Robb. This is about your image; this is about what you've done in the community." I knew I couldn't do that. It felt like idolatry. The real problem with propaganda is that it creates allegiance towards something that's not God. And I am a follower of Jesus. So I struggled with that.

When I was discerning whether to run or not, a long series of conversations with others helped me understand that it came down to two things. Could I run a campaign where I could be honest about my limits and the limits of political power? I brought that commitment into the campaign, but my campaign team said, "Do not ever talk about that." Without telling my campaign team, I wrote a local news blog entitled, "I'm going to disappoint you." I tried to communicate, "you are projecting on me many hopes.

   3. Ellul, *Présence au monde modern*, 19. Robb's translation.
   4. Ellul, *Technological Society*, 375.

You are projecting on me your desires, and there's no way I can fulfill those things." The blog was an intentional act. All I offered was the ability to try to bring people together, to try to work together to solve some of the issues. My campaign team got quite upset.

Mark: With the campaign team, was it a one-time disagreement, or ongoing conflict?

Robb: It was ongoing conflict, but not about everything. For instance, I made a commitment during the campaign, that my political career begins and ends in Davis. So I am committed to localism. I'm committed to this bioregion. I'm committed to naming the giftedness of the people in this town and drawing on that giftedness to solve our problems. I laid that out and I said, "I will not seek higher office." My campaign team was okay with that.

Mark: Let's return to your comment about confession for collective sin. Can you give an example of how you do that?

Robb: I am frequently asked to speak at events. Recently I spoke at a demonstration against Bakken crude oil coming through our town by rail. It is very volatile and there have been railroad accidents and explosions in other places, killing many people and causing significant environmental destruction. What I mean by public confession is standing in front of a group of environmental activists and saying, "You know the oil company is not going to the Bakken formation to make our lives miserable. The oil-producing company is not going to the Bakken shale to give us heartache, or to challenge our goal of local control of land use. They're going to the Bakken shale because we're telling them to. We're asking them, we're begging them, our society, our lifestyles are drenched in oil. That's why they're going." Now, that's my public confession of my participation in systemic sin. We're raping Canada's timber to build houses in California. We've despoiled the Ecuadorian rainforests to drive our cars. We need to say that; we need to acknowledge our role in the systemic.

Mark: Ellul wrote: "The first great fact which emerges from our civilization is that today everything has become 'means.' There is no longer an 'end'; we do not know whither we are going. We have forgotten our collective ends, and we possess great means: we set huge machines in motion in order to arrive nowhere."[5] How have you observed this?

5. Ellul, *Presence of the Kingdom*, 63.

Robb: Two months after I was elected, an MRAP, Mine-Resistant Armored Personnel Carrier, arrived in our town. It looks like a tank without a turret. It was surplus military equipment sent by the U.S. Government at the request of our police department. I asked, "We need a tank?" And the police said, "Yes. We need it for lone shooter events were somebody's hiding and shooting. We need it in case of a disaster. We need it in case there's a riot."

Means and ends. The day it arrived, the first thing that came into my mind was, "Means and ends." What did Ellul say about means and ends? Now let's think about this vehicle, the MRAP. It has an end. It was developed for a reason. It was developed for one very specific reason. It can carry large numbers of soldiers down a flat Iraqi road, have an explosive device go off underneath it, and preserve the lives of the people inside. When the war was over, the U.S. Government needed to do something with it, and so it committed to sending these MRAP's to every community that wanted one in the United States, no strings attached. A vehicle worth $750,000 each.

And our police are saying to me, "We need it. We need it." So I challenged them, and I said, "What's the concern? Security, right? We need it for our security." And we did Town Hall meetings, and people came and said, "We need it for our security." That's the end that we're trying to achieve, security.

So, I asked the police in public meetings, "What's the security threat?" They said two things. Think of this through the lens of Ellul. Everything is becoming means. We've forgotten the ends. We have a machine that's created for certain ends, now this machine, this means, is coming to a community and what we're trying to do is find an end that justifies this means so that we can keep it. We "create" ends to justify its continued use. But it's an instrument of power and control. And so, the police said, "Well, we have drug deals going down in our town, and the drug dealers are stealing each other's stashes, and they get into gun battles with each other, and we need it in case we're going in to arrest the drug dealers because they're heavily armed."

Now think about that in terms of ends. The first question was, "Who's buying the drugs?" And the police turned to me and said, "We have a heroin problem among our young people and a methamphetamine problem among our middle-aged population." So how do we respond to this problem? We're going to address addiction with an MRAP. We are trying to achieve certain ends (reduction in drug sales) by focusing on the wrong means. We should be looking at the causes of addiction, not stopping drug sales caused by it with an MRAP.

The second one is even more telling. It gave me chills and I hope it gives you chills too. The assistant chief of police came to me separately, and said, "Robb, we have legitimate concerns. There are former military people in this community who are tactically trained. They're trained in police tactics, and they know how to counter us, and by the way Robb—some of these folks have PTSD. If they get guns in their hands, it's very difficult for us to deal with them."

Means and ends, right? We go off to Iraq. We wage war. Men come back with PTSD, tactically trained. And the way we deal with them is an MRAP so that we can take them out? This is the way we're dealing with the problems in our community? With an MRAP? So we voted to get rid of it. It felt significant, but the Department of Defense sent it ten miles north to the city of Woodland. We were the laughingstock of the neighborhood.

The big blowback came a few weeks later though and relates to another insight from Ellul. In the film, "The Betrayal of Technology" he said, "Technique will not tolerate (or accept) any judgment passed on it. In other words, technicians do not easily tolerate people expressing an ethical or moral judgment on what they do. . . . Technique does not accept judgment." I find it very interesting that he started by saying, "*la technique*." It shows me that technique is a spiritual power. In addition to the technicians, there is *la technique*, which is the Power. The blowback we got, which was severe, and I almost thought I was going to be recalled, was that we were accused of compromising the security of our city. I sat with the police and the police said, "*We* are the experts. *We* understand security. *You* are a politician, you do not know about security, you've taken a tool of security out of our hands." I said to them in a public meeting, "The problem I have with the MRAP is that it is a symbol." It is a symbol of the most destructive military force that the world has ever known, and we're bringing that into our community."

Most politicians don't want to talk about ends, because a lot of times the ends that they're working towards are hidden. They're not the ends that they say publicly. Push them on ends. Push them. Push them. The other thing is that we do have, in every bureaucracy, we have people who are enamored with means who will look for ends to which the means can be applied. It is means in search of ends.

Mark: In what ways have you personally felt challenged in relation to these themes we have been talking about, and what have you done in response?

Robb: People don't corrupt you overtly. They do it this way: "Man, you're amazing. You know if you push this it'll pass because people respect you. Could you push it?" It's subtle. Every single day there is the temptation to use power. For instance, I want to work on restorative justice with youth. One day I pick up the newspaper and it says, "Robb Davis led the initiative on restorative justice." I read it and think, "Actually, no I didn't. There were ten of us in the room." I have a choice at that point. Am I going to go correct the paper and say, "Actually there were ten of us in the room, and I didn't lead anything." or, am I going to let that go? Most people would say, "Let it go. Let it go. People will follow you, and you'll be able to move much more quickly."

What happens when you start listening to those voices? Here's what happens: The goal is restorative justice. That is the end that you want to achieve. You start believing that what I need to do is accumulate a little more of that status and power so that I can be better at bringing change. It can lead to two things. First, I can use the positive end, restorative justice, to justify means inconsistent with restorative justice itself and, for me, importantly, inconsistent with the way of Jesus. Second, with increased emphasis on the means to achieve power, eventually the original end of implementing the practice of restorative justice can get lost. Achieving power becomes the true end—even if not the acknowledged one.

Therefore, I must re-orient regularly. I so easily get pulled off track. As part of that re-orientation I have had to do things like go before people and say, "You know what, I should've spoken up earlier, I had nothing to do with that. I didn't do anything about that. I can't take any credit for that."

Mark: As you point out, to make effectiveness the supreme goal can become problematic, yet you do seek to be effective, correct? As you state, you desire to see an increased practice of restorative justice. You want to be effective in that.

Robb: Yes, we can't live without some commitment to effectiveness. The problem is making effectiveness or efficiency the supreme goal that drives and determines everything. I have found it is of utmost importance to have made premediated commitments. For instance, like Ellul I am committed to not use violence. Without that commitment, if violence appeared to be required to achieve a goal, I might too easily succumb to the ends justifying that means—the means of violence. Ellul has certainly been a key influence in helping me, as a follower of Jesus, determine what my pre-commitments are—things I will not do in spite of what efficiency may demand or promise. This is not to say I am always faithful. As I just said, re-orientation is a constant necessity.

Interview Part Two: A Year Later

Mark: When we talked before you were about to start as mayor. In relation to what we discussed last year what new insights do you have after a year as mayor? Could you share some experiences that further illuminate points you made or that have led you to think about Ellul's analysis in new ways?

Robb: Today, I am even more convinced that one of the great dangers that we have is that we use propaganda to sell the image of the good leader, the strong leader. What I have seen nationally, since the election and even before, is that this use of propaganda has increased. The only way to confront it is to be confessional and eschew the acts that would help mobilize power toward oneself. Therefore, in my writing and my speaking I have sought to be confessional; to be vulnerable and acknowledge the brokenness.

Colleagues on the city council quite frequently ask, "Robb, why do you take the blame, why do you take responsibility? Why do you constantly acknowledge we could have done better?" My attitude is that we take responsibility not as a pro-forma thing. We acknowledge our failings because they are real, they are true. People appreciate confession, but I do not think it mobilizes allegiance.

Mark: One cannot accumulate power from a stance of confession and therefore rather than confession politicians use propaganda.

Robb: Correct. Those advising people in power tell them to never admit mistakes or acknowledge harms. In effect, they are telling them, never do anything restorative. I think it is my responsibility to seek restoration. It is important for me to understand the harms I have caused and acknowledge them.

Mark: As Ellul wrote, "[A] political leader will not make true decisions as long as appearances will save him."[6] Last year you gave an example of speaking a confession for collective sin, can you give an example of personal confession?

Robb: Recently I apologized for not moving forward an agenda on downtown parking that I lost energy on. Similarly, I have confessed publicly to running out of energy to advance some of the goals we have on homelessness. I did not remain focused on it and other things came up. I would say that three or four times a month I apologize for missing something, not getting it right, or for upsetting people in ways I did not anticipate or intend.

---

6. Ellul, *Political Illusion*, 32.

Mark: Some of what you have said leads me to think of these lines from Ellul's *Presence of the Kingdom*, "The Christian is a visible sign of the new covenant which God has made with this world in Jesus Christ. But it is essential that the Christian should really *be* this sign. . . . The fact that Christians *are*, in their lives, the 'salt of the earth,' does far more for the preservation of the world than any external action."[7]

Robb: Yes, I am more than ever struck, in this day and this time, for the need to be human with each other. The processes that have been unleashed in our nation and the divisions that exist have led to people feeling dislocated and disoriented.

There was a handicapped man who uses a wheelchair. He was quite frustrated by many accessibility issues in the city. We met one evening at a downtown coffee shop. He had two full pages of notes of things that he wanted to tell me. He took me to a couple venues where there were access problems. Then, showing me where he parked, he shared his frustration with people who have handicapped placards who are not really handicapped. He wanted me to see what was involved in him getting out of his van, how important the handicap parking space was. He said to me, "Would you be willing to get into a chair and see some things from the perspective of someone in a wheel chair?" He pointed me to another wheelchair in his van. We then wheeled off down the street. I followed him for about forty-five minutes. He wanted to show me how difficult it was to cross some of the major streets. And it was quite difficult—eye opening. When we got back to his van, he held up the list and said, "I had this list I was going to give you, but what really matters is that you listened to me."

Recently there was a hate crime incident at the mosque, not as serious as an earlier one, but still deeply troubling. Immediately, in response, people and groups sent out messages to come to the mosque that evening to demonstrate solidarity. I went, not knowing what was to be expected, and ended up spending the whole evening there. I had dinner with Moslem people, talked about this and that—not much about the incident itself. Several of them thanked me for coming. They told me what mattered was that I came, my presence.

Police intervened in a drunken brawl; there were some racial factors involved and Black Lives Matter folks started coming to council meetings demanding we take action. They were quite hostile. Unfortunately, they personalized their anger at me, the mayor, for not acting. One meeting it was so hurtful to me, I felt so bad, I suspended the meeting for 15 minutes. I went

---

7. Ellul, *Presence of the Kingdom*, 9–10.

outside to be alone because it was so hard to hear these things coming at us. Yet, later I gave them my business card and asked them to get in touch with me. I have met them; we have sat down and had coffee together. We have explored what their concerns are and my constraints.

Mark: These examples not only point to what Ellul wrote in those lines in *Presence of the Kingdom*, they also illustrate a contrast he makes in *The Political Illusion* between true democracy and what passes for democracy today. A truly democratic person must have "absolute respect for adversary, fellow man, minorities: respect that has nothing to do with liberalism . . . nothing to do with tolerance." Rather it is true dialogue that recognizes differences and commonalities.[8] During your campaign you stated you would not seek a higher office. Earlier this year you announced you were not seeking re-election. What changes have you observed since making that decision?

Robb: An important reason I am not running again is that our electoral system, even at the local level, requires us to put six to eight months into campaigning. When you have a full-time job, which all of us on the city council do, and you have council business, and you have to campaign—something has to give. I feared city business would lose out. I am here to govern. I committed to govern for four years. I want to finish the job I set out to do.

I now speak with more freedom. Propaganda is also about what we hide, what we do not say. Being in a position where I can speak frankly and not worry about the outcomes allows me to shine a light on things that people are very uncomfortable having a light shone on. I have written and spoken in public more directly—whether about the criminal justice system, addiction, homelessness. I have upset some people because I have spoken more directly than people are comfortable with. I am not setting out to shock people, but when you do not have that burden of trying to keep everyone happy or not burn bridges you can speak more freely.

One example, I could give many; I wrote a piece for the newspaper on hate and hospitality.[9] I used vignettes from when I lived in the Muslim world where I learned about hospitality and contrast that with the hate crime incident at the mosque here in Davis. I am not sure I would have written that if I was running again. It is a little risky to say that I learned hospitality from Muslims, and then state that now in my hometown they as a group are being treated inhospitably.

---

8. Ellul, *Political Illusion*, 236–37.
9. Davis, "Hate and Hospitality."

Mark: What connections do you see between Ellul's thought and what you have just described?

Robb: I see it as being consistent with Ellul's concern about efficiency being the ultimate criteria. What I keep going back to in Ellul is that we are called to faithfulness in this local space. It is not efficiency that matters; it is holding the complexity of the world around us and in the day-to-day challenging it by being real humans. What I have said above has emphasized being with people, walking with them in the journey they are on to try to find justice and reconciliation. Rather than having to scheme and move forward an efficient plan of action, Ellul challenges me to be human, be with people. Technique is profoundly dehumanizing. Human beings become secondary. When we focus on power, image and legacy we become enamored with means and forget about the ends.

Bibliography

Davis, Robb. "Hate and Hospitality." *Davis Vanguard*, July 6, 2017. http://www.davisvanguard.org/2017/07/hate-and-hospitality/.
Ellul, Jacques. *The Political Illusion*. Translated by Konrad Kellen. New York: Knopf, 1967.
———. *Présence au monde modern*. Geneva: Roulet, 1948.
———. *The Presence of the Kingdom*. Translated by Olive Wyon. Philadelphia: Westminster, 1951.
———. *The Technological Society*. Translated by John Wilkinson. New York: Knopf, 1964.
Ellul, Jacques, and Madeleine Garrigou-Lagrange. *In Season, Out of Season: An Introduction to the Thought of Jacques Ellul*. San Francisco: Harper & Row, 1982.
Ellul, Jacques, and Patrick Troude-Chastenet. *Jacques Ellul on Politics, Technology, and Christianity: Conversations with Patrick Troude-Chastenet*. Eugene, OR: Wipf & Stock, 2005.

14

Jacques Ellul and Thai Politics since 2000

BY Virginia Landgraf

> *Virginia Landgraf is an indexer-analyst at the American Theological Library Association. She wrote her PhD dissertation, "Abstract Power and the God of Love: A Critical Assessment of the Place of Institutions in Jacques Ellul's Anthropology of Dialectical Relationships," under Max Stackhouse at Princeton Theological Seminary. She has presented and published on Ellul's interpretation of Sennacherib in Second Kings and on readings of the Ten Commandments that can be derived from Ellul's work. She has contributed repeatedly to the* Ellul Forum *and IJES conferences and is a member of the IJES board of directors. For the 100th anniversary of Ellul's birth she contributed an essay to* Réforme *on the reception of Ellul in the USA. In 1988–1990 she served with the U.S. Peace Corps in Thailand and spent additional time there studying Thai music.*

* * * *

Jacques Ellul on Societal Resilience, its Modern Enemies, and the Possible Role of Law

A LITTLE-KNOWN SECTION IN Jacques Ellul's *The Political Illusion* sketches out a view of a healthy society which contrasts with the depictions of how efficiency crowds out other ends which dominate his sociological work. Ellul presumes that a polity or society will face challenges that threaten its viability, whether they be rivalry from other societies or trends within the society and its physical substrate itself (economic, demographic, environmental, etc.). A resilient society is one that is successfully able to meet these challenges, which cannot be exhaustively predicted in advance. Using a model from information theory, he compares societies to systems that have more or less capacity to receive new information and hence redirect themselves when these various challenges arise. A system that is completely fixed or a system that is completely random cannot receive new information in a way that will affect its self-direction. Ellul sees two prerequisites for a polity to avoid either of these dead ends: tensions between groups in a polity and recognition of a common measure belonging to all members of a polity, regardless of group membership. Without tension, there is the danger of a fixed, dead system; without a common measure, there is the danger of a chaotic war of all against all. [1]

Ellul sees the usurpation of politics by the imperative of efficiency as a threat to both the maintenance of healthy tensions and the recognition of commonalities among members of all groups. Technique, invading all areas of life, not just politics, could lead to a totally conformist society, in which everybody is so consumed with their technical jobs and escapist entertainments that they give no thought to long-term challenges to their way of life.[2] On the other hand, the manipulation of people's sense of identity in the course of the process used to gain power in modern states threatens the recognition of a common measure with one's political opponents. One may be encouraged to label adherents of an opposing political ideology as having stepped outside the bounds of that polity by virtue of their political beliefs (cf. the hunting down of Communists by the "House Committee on Un-American Activities" in the 1950s).

As a Christian, Ellul believed that it is ultimately the Word of God that is the new information challenging the imperative to seek self-preservation by political or economic power. This is his interpretation of the white horse fighting the red, black, and pale horses in Rev 6:1–8 and 19:11–16. Without this challenge, political and economic power (the red and black horses respectively) join the pale horse (disease and decay) in meting out death to the world.[3] However, *The Political Illusion* uses arguments meant to be convincing to people of any faith or none. Given the apparently observable, although perhaps vague and general, criteria of "tensions between groups" and "recognition of a common measure" to assess the potential resilience of a polity, one may ask whether there are factors within a polity or society's governance or common life which may help or hinder the maintenance of these criteria.

1. Ellul, *Political Illusion*, 206–23.
2. Ellul, *Technological Society*, 375–82, 400–402.
3. Ellul, *Apocalypse*, 108–10, 147–56.

Ellul does not see the enshrinement of democracy in constitutional law as a sufficient or necessary criterion for a polity that allows healthy tensions between groups and recognizes a common measure among all its members. He sees the liberal state, despite its democratic pretensions, as captive to an ideology of material well-being that blunts its ability to receive information from those on the underside of that ideology.[4] He thinks that his contemporaries during the Cold War placed too much of a burden on institutional arrangements (socialism or capitalism) as the remedy for social ills, neglecting individual responsibility.[5] His advocacy of nonviolent anarchy was a reflection of his belief that in his context, the state and its power had taken over society. "Every modern state is totalitarian," he thought,[6] and to oppose such totalitarianism one had to oppose not only the state but institutions that (in the context of the imperative to technical efficiency) played into the state's power by dealing with people as abstractions rather than fostering face-to-face relationships.

On the other hand, Ellul's writings on "dialogue with the sovereign," recommending that Christians challenge those in power on their own terms, imply that there is a positive role for explicit ideals in perhaps letting new information be received. He says that if those in power (whether they be an individual, such as a monarch, or one's neighbors as constituents of the apparently sovereign people) say that they are for, for example, justice, that Christians should challenge them when their policies do not bring justice.[7] "Dialogue with the sovereign" is distinct from advocating that the state proclaim itself to be Christian. Ellul strongly opposes any such move. He sees a legitimate role for political authority (as distinct from an abstract state) to be a wise manager of common patrimony. Such a limited role does not require religious pretensions.[8] In fact, he sees the religious pretensions of states and political actors as reinforcing state power and suppressing genuine dialogue, either in terms of absolute attachments to political identities or personality cults of leaders.[9]

Despite his belief that constitutional law does not guarantee a polity where people are able to carefully consider alternative courses of action and make their voices heard, Ellul was willing to challenge the state on the basis

---

4. Ellul, *Ethics of Freedom*, 370.
5. Ellul, *Money and Power*, 15–23.
6. Ellul, *Ethics of Freedom*, 396.
7. Ellul, *Ethics of Freedom*, 392.
8. Ellul, "Rappels et réflexions sur une théologie de l'État," 173.
9. Ellul, *New Demons*, 166–202.

of its laws in his advocacy for environmental protection.[10] This willingness qualifies Ellul's statements elsewhere, noted by Andrew Goddard, that law was not far from the demonic.[11] The ideal and reality (or lack thereof) of the rule of law deserve a closer look in the pursuit of a society where there are healthy tensions between groups amid the recognition of a common membership for all. Ellul saw law, like many other areas of life, as an arena in which the pursuit of technical perfection was undercutting respect for persons and discernment of distinctive circumstances. A technical system of law will tolerate injustice as a means to maintain order.[12] And even in societies that claim the rule of law, that same law and the processes of its administration may be instruments of oppression for some members of that society, as Michelle Alexander has pointed out with respect to the disproportionate incarceration of African-Americans in the United States. [13]

Legal codes can be used to enrich rulers at the expense of the common good or manage underlings who do not share in the rights afforded by the law. However, in theory, the rule of law acknowledges that even the rulers are under the law. In this sense, the rule of law should be a limit on power, as pointed out by theologian and lawyer David McIlroy.[14] And it should provide a common measure among members of a polity. With respect to the other of the two criteria for societal resilience—tension—a system of law that allows for tension would be modest and sufficiently flexible to allow for new circumstances to be taken into account. It would not attempt to manage citizens' lives so comprehensively that the only way for tension to arise would be for the law to be broken. To place exhaustively all of a society's members' activities under legal direction would be to forestall the receipt of new information, however futile such an attempt might be.

The remainder of this paper will be devoted to a concrete study of a particular polity, the kingdom of Thailand. Thailand is interesting to examine in light of Ellul's ideas about societal resilience and the threats to it because it is neither a strictly liberal state—although it is a constitutional monarchy, the King is generally acknowledged to have played a role in politics at least from the 1970s on—nor a state proclaiming a well-known totalitarian ideology. It is also an interesting case because most Thais are Theravada Buddhists. Ellul is known for criticizing Christianity for

---

10. Ellul, *Anarchy and Christianity*, 16.
11. Goddard, *Living the Word*, 255.
12. Ellul, *Technological Society*, 293–300.
13. Alexander, *New Jim Crow*.
14. McIlroy, "How Is the Rule of Law a Limit on Power?," 38–44.

making supposedly Christian institutions or experiences sacred[15] rather than being open to the newness of the Word of God, which should enable one to be free from all ideologies.[16] Buddhism, with its doctrine that no thing has a self that should be grasped at or clung to, has in theory a similar desacralizing potential.

My purpose is to analyze Thai politics from the point of view of Ellul's ideas about societal resilience, not to give an exhaustive accounting of events. The ideal situation (although Ellul would be hesitant to say that an ideal situation can be fixed) would be one in which those in power are subordinate to the rule of law; where there is freedom of expression and investigation enabling citizens to examine and question existing policies or the acts of the powerful; and where enduring truths beyond immediate political exigencies can affect discourse. When these conditions are not in place, it becomes more difficult to maintain the balance between intergroup tension and recognition of a common measure among all.

My conclusion will be that none of the rival parties involved have operated within frameworks where those in power are fully subordinate to the rule of law and peaceful political dissent can proceed undisturbed. The severe lèse-majesté laws make it difficult to discuss good or bad aspects of particular royal policies. In addition, although Buddhism has doctrines that should encourage the questioning of abuses of power, politicians have also used elements of Buddhist cosmologies to legitimate power.

Major Events and Factors in Thai Politics since 2000

Thai politics since around 2000 can be seen as a drama between an electoral game-changer, Thaksin Shinawatra, and his enemies, with the monarchy in the background as an ever-present player facing the inevitable challenge of royal succession. Thaksin,[17] who became prime minister in 2001, was the first Thai prime minister whose party managed to win an absolute electoral majority in parliament. Although he was a telecommunications tycoon, he built his electoral base on residents of the poorer north and northeast of the country. Governments under him and his followers supported policies that bolstered the safety net for the poor, slashing the price paid by consumers for certain kinds of health care and rail transportation and guaranteeing a

---

15. Ellul, *Subversion of Christianity*, 31–32, 62–68.
16. Ellul, *Ethics of Freedom*, 334–35.
17. Thai people are traditionally referred to by their first names, which are their given names, rather than their surnames, which are their family names.

minimum price for rice. They also favored capitalistic economic development and greater integration of Thailand into the international economy.[18]

However, Thaksin's increasing power did not sit well with most of the Bangkok elite and some of the established politicians. He was ousted in a 2006 coup while on state visits abroad and went into exile. Subsequently he was banned from politics for five years and convicted of conflict of interest, and his family was stripped of $1.4 billion in assets.[19] Following the 2006 coup, Thaksin's followers formed another political party and continued to win elections once civilian government was reinstated. Opponents of Thaksin came to see electoral politics as an insufficient means to rid the country of his followers or policies. Groups of demonstrators wearing yellow, the color of the king, protested pro-Thaksin governments, shutting down Bangkok's major airport in 2008.[20] Pro-Thaksin demonstrators also organized themselves, taking red as their color. The rivalry between "Yellow Shirts" and "Red Shirts," sometimes becoming violent, continued into the 2010s.

Near the end of 2013, a pro-Thaksin government under Yingluck Shinawatra, Thaksin's sister, had been in power since 2011. However, anti-Thaksin demonstrators led by former deputy prime minister Suthep Thaugsuban threatened to "shut down Bangkok" in January. Yingluck called a snap election for early February, putting her government into caretaker status. The largest anti-Thaksin political party, the Democratic Party, refused to contest the elections.[21] Suthep's demonstrations continued as planned, featuring a combination of speeches attacking Thaksin's family and allies for corruption and civil servants reading statements saying that the caretaker government had no right to interfere with their freedom of speech and assembly.[22] The demonstrators called for "reform before elections" that would put "good people" in to administer the government (and, presumably, prevent bad people from hijacking the electoral process).[23]

The elections of February were held, but anti-Thaksin demonstrators were able to interfere with enough polling places that they were declared invalid.[24] A lawsuit accusing Yingluck of unfairly using her office to benefit a relative was brought and moved swiftly through the courts, in contrast to

18. Rock, *Dictators, Democrats, and Development*, 228–33.
19. BBC News, "Profile: Thaksin Shinawatra."
20. MacKinnon, "Rioting Protesters"; MacKinnon and Walker, "First Flight."
21. Fuller, "Thai Opposition Party."
22. Demonstrations broadcast on Blue Sky Channel TV in Thailand, seen by the author there in late January 2014. Comparable footage is included in Blue Sky Channel TV (Thailand), "BLUE SKY LIVE International."
23. Merisa, "Suthep's Romantic Tale."
24. Hawley, "Thailand Constitutional Court."

a case of murder involving Suthep, which had languished for four years and was eventually dismissed.[25] In May 2014, martial law was declared, and a military coup followed a few days later.[26] Many activists were arrested, and some were tortured, being released only if they would sign a pledge not to engage in politics.[27] The ensuing dictatorship, under Prayuth Chan-o-cha, has not yet relinquished rule as of November 2017, has rewritten the constitution to restrict the effect of elections on the legislature, has forbidden criticism of the junta, and has commissioned a new line of textbooks that expunges Thaksin's name from the history books.[28]

Some observers believe that the Thai monarchy supported the anti-Thaksin demonstrations and coups. They see Thaksin as a figure who threatened the charisma of the king.[29] As the drama between politicians was unfolding, the monarchy was facing the challenge of King Bhumibol Adulyadej (Rama IX)'s long physical decline (born in 1927, he died in 2016) and the fact that the crown prince, now King Maha Vajiralongkorn (Rama X), did not enjoy the popularity or image of public-spiritedness that his father had. King Bhumibol was depicted as a king zealous for his people's welfare, eager to encourage development projects around the kingdom. By contrast, Prince Vajiralongkorn had the image of a playboy who had been married several times and had disinherited several children.[30]

Thai Politics since 2000 vis-à-vis the Rule of Law and Civil Liberties

Amid this drama, violations of the rule of law and restrictions on civil liberties have occurred at several levels. When Thaksin was in power before the 2006 coup, reports from international human rights activists were concerned about the use of extrajudicial killings and restrictions on the freedom of the press and assembly.[31] By contrast, under the last pro-Thaksin government before the 2014 coup, Suthep's anti-Thaksin speeches were broadcast on widely available private television networks. Whether that apparent press freedom was an indication that pro-Thaksin governments could learn from

---

25. Fuller, "Thai Prime Minister"; Tan, "Murder Charges Dismissed."
26. BBC News, "Thailand Military."
27. Asian Human Rights Commission, "THAILAND."
28. Human Rights Watch, "Thailand"; Fuller, "Loved and Hated."
29. Jory, *Thailand's Theory of Monarchy*, 187.
30. BBC News, "Profile: Thailand's New King Vajiralongkorn."
31. Human Rights Watch, *World Report 2005*, 326–32.

previous mistakes ended up being a moot point.³² The interaction between the demonstrators and those they opposed was not eventually characterized by dialogue, but of might makes right, as the demonstrators, combined with the electoral commission, managed to suppress the election, and the junta has governed via suppression of civil liberties as enumerated above.

Furthermore, every time there is a coup, the rule of law is violated. Those who have seized power are allowed to rewrite the rules. Since the coup that installed the constitutional monarchy in 1932, there have been twelve coups,³³ and they conform to a pattern. Conflict develops under the old constitution that seems unresolvable by parliamentary process. The military intervenes, suspends the constitution, and presents their government to the king. The coup-makers do not have to face consequences for breaking the previous constitution. Eventually a new constitution is written, and there is a return to elections, civilian rule, and parliamentary process, but only until the next irresolvable conflict, when a new coup occurs and the cycle begins again.³⁴

The monarchy is crucially important in assessing both the rule of law and the state of civil liberties in Thailand. Strict lèse-majesté laws prescribe a three to fifteen-year jail sentence for defaming the king or royal family.³⁵ Despite King Rama IX's 2005 birthday speech in which he said that he could be criticized,³⁶ the lèse-majesté laws have been enforced by every government, pro- or anti-Thaksin, since that time. By forbidding defamation of the royal family, royal projects, or past kings, these laws discourage citizens from discussing the purposes of the monarchy from any point of view other than that it should be maintained and revered. Even Buddhist social critic Sulak Sivaraksa, who wrote laudatory poetry for King Rama IX, has been accused of lèse-majesté multiple times.³⁷ As of November 2017 a case is pending against him for declaring that the traditional account of King Naresuan's victory over the Burmese king in a sixteenth-century elephant

---

32. For a characterization of the anti-Thaksin protesters as wanting "clean" government and the pro-Thaksin supporters as wanting "responsive" government (as well as other dimensions of analysis), see Saxer, "How Thailand's Middle Class."

33. Taylor and Kaphle, "Thailand's Army."

34. Bangkok Pundit, "Vicious Cycle." The diagram, putting the terms "coup," "military rule," "constitution," "election," "parliamentary process," "conflict," and "crisis" into a circle, with "coup" at the top, is believed to have been published originally in Chai-Anan, *Thai Young Turks*, and reprinted in Likhit, *Thai Politics*.

35. Streckfuss, *Truth on Trial in Thailand*, 109.

36. Bhumibol Adulyadej, "His Majesty the King's Birthday Speech."

37. Streckfuss, *Truth on Trial*, 192–93, 202–3.

battle may be a legend and not fact.[38] The net effect is that an institution whose assets have been estimated from $30 billion to nearly $60 billion[39] and exerts considerable moral and political influence is declared off-limits for criticism, even by those who are otherwise sympathetic to the idea that a wise Buddhist monarch can exert positive influence on a kingdom.

Ironies of Buddhism and Politics in Thailand

It is ironic that a kingdom where Buddhism is the majority religion has such a sacralizing policy toward its monarchy. According to Buddhist doctrine, all things are anattā, meaning that they have no self or essence that should be grasped at,[40] and anicca, meaning that they are impermanent, in a constant state of flux and becoming.[41] Scholar of Thai Buddhism Peter Jackson has noted the contrasts between the Brahmanic devaraja doctrine of kingship, where the king was an avatar of the divine, and the Buddhist dhammaraja doctrine, where the king was supposed to be guided by Buddhist moral law. He sees the development of the symbolism and mythology of the Thai monarchy over the past half-century as akin more to the god-king than the dharma-king tradition.[42]

King Rama IX's birthday speech of 2005 seemed more in keeping with the dharma-king tradition. There he said that if he could not be criticized, he would not be human. He urged people to take responsibility for their actions and cultivate awareness before they speak. He pointed out his pardons of people accused of lèse-majesté.[43] Analysts have interpreted this speech in multiple ways. What one might call the "royal apologist" tradition interprets it as meaning that the king has dharma and compassion.[44] On the other hand, the speech has been interpreted as directed at then-Prime Minister Thaksin's difficulties in handling criticism.[45] One does not have to settle on

---

38. Cochrane, "Buddhist Faces 15 Years in Jail."

39. Pavin, "Very Wealthy Monarch," has a figure of nearly $60 billion for the wealth of the Crown Property Bureau; BBC News, "Profile: Thailand's New King Vajiralongkorn," has a figure of $30–40 billion.

40. Buddhadāsa Bhikkhu, *Buddha's Doctrine of Anattā*, 15.

41. Knitter, *Without Buddha*, 10.

42. Jackson, "Virtual Divinity," 47.

43. Bhumibol Adulyadej, "His Majesty the King's Birthday Speech (Thailand)."

44. A video of the relevant section of the speech in the Thai audio with English subtitles emphasizes that the king has compassion and dharma and cannot be destroyed by ignorance (e.g., foreigners who deface posters of the king). Phraezila, "'If They Get Sent to Prison.'"

45. Connors, "King Can Do Wrong."

one interpretation or another to take the speech at face value and use it for the purposes of "dialogue with the sovereign." Buddhism recognizes that human beings are going to die and can make mistakes. It is not clear that the lèse-majesté laws have taken that acknowledgment to heart.

Also, Buddhism's emphasis on profound insight into reality may help people focus on sources of tension that are not life-destroying. Instead of politics based on branding and group identities that exclude the other, one might imagine politics based on local ecological and cultural knowledge and the recognition of limits. There would be tensions, because not every option could be pursued. A related concern is to pursue ways of acquiring and exercising power that do not violate the precepts against lying or intoxication: not to attempt to distort facts to remain in power, become attached to power or its advantages, or be afraid of losing power.[46] The call to extend loving-kindness to all beings[47] applies even to one's political enemies. Some ways of telling the legendary previous lives of the Buddha (jātakas) depict Buddhist wisdom as interrupting vicious circles of cruelty or power-seeking.[48]

However, doctrines that teach self-surrender, detachment, loving-kindness, unflinching observation of reality, and the interruption of vicious circles of power are not the only ideas at play in contemporary understandings of Buddhist kingship. Andrew MacGregor Marshall has found that popular understandings of karma in Thailand tend to justify the position of those in power: those who have power must have done something in previous lives to deserve it.[49] Patrick Jory has pointed out that the concept of barami, moral virtue that accumulates over successive lives and is especially expressed through generosity, is an important attribute of the king. Jātaka tales may be interpreted not as expressing the disruptive effect of Buddhist wisdom on vicious circles of power but as confirming the king's generosity. Thaksin could be seen as a rival to the king precisely because his policies seemed to be expressing or accumulating barami.[50]

Over the course of the reign of Rama IX a visual culture of royalist propaganda developed, not only showing the king visiting development projects, but the king and sometimes the queen posing in formal clothes, looking forward beatifically, often with the words song pra charoen inscribed

---

46. The fifth precept has been rendered as "I observe the precept of not taking intoxicating things which are conducive to bewilderment." Vorasak, *Pāli Recitations*, 75.

47. "Karanīyametta Sutta," Vorasak, *Pāli Recitations*, 64.

48. See, for example, "Nang Kunthalakesi," in Po, *Chumnum Nitan Chadok*, 29–52.

49. Marshall, *Kingdom in Crisis*, 29.

50. Jory, *Thailand's Theory of Monarchy*, 16–19, 180–88.

somewhere on the portrait.[51] Song pra charoen can be roughly translated as "Long live the king," but "may blessedly prosper" is a more literal translation, since there is no noun in the expression. With propaganda providing something to worship and lèse-majesté laws deterring criticism, the Thai monarchy takes on characteristics of the personality cults that Ellul criticizes among "political religions" in The New Demons—people who are presumed to be always right and are objects of worship—rather than functioning as an independent voice above the need to compete for power.[52]

Conclusion

Many of the policies of Thailand's current military dictatorship, including the lèse-majesté laws which predated it, attempt to maintain order by suppressing difference or questioning. In Ellulian terms, they seek a common measure by suppressing tension and thus risk being unable to take in new information to meet societal challenges. While no government among those examined here has been immune to this temptation, the forbidding of criticism puts the onus on those who are legally beyond criticism (currently, the junta and the monarchy) to be infallible. This burden seems untenable. It also presumes an epistemic privilege that sits awkwardly with Buddhist doctrines of anattā and anicca.

Ellul might say that the suppression of civil liberties by Thai governments and the distortion of the Thai monarchy's possible purpose under a virtuous monarch is a reflection of the fact that "every modern state is totalitarian." Commentators have noted that the Thai monarchy's propaganda apparatus and power to rule do not exist in a vacuum but have developed within the contexts of playing off or cooperating with rival colonial or neocolonial powers.[53] Ellul might also question whether Buddhism can truly bring a new word from outside the vicious circles of power. As a Christian theologian, I do not think that I have to settle that question in advance. I trust in God's ability to speak, whether through Christians or others. My selection of doctrines from Theravada Buddhism is meant to be both an appreciative and a critical inquiry, but not exhaustive or metaphysically conclusive. I have lifted up Buddhist doctrines that I think can be helpful in the fight against political idolatry and criticized those that seem less helpful.

---

51. Examples of royal iconography can be seen in summaries of King Rama IX's life such as Head, "How King Bhumibol," and Campbell, "Life in Pictures."

52. Ellul, *New Demons*, 170–78.

53. Jory, *Thailand's Theory of Monarchy*, 184–85.

This analysis is also not meant to imply that the societal dysfunction I describe in Thailand does not happen in Western countries as well. Politicians are marketed as rock stars, regardless of the substance of their policies. A female scion of a political dynasty practices explicitly inclusive propaganda[54] as opposed to her opponents' divisiveness, but questions remain whether her policies are fiscally sustainable and whether her family isn't illicitly enriching itself as a result of being in power. Angry groups longing for a purity beyond the current régime disrupt politics as usual by both electoral and extra-electoral means. People who complain loudly about corruption before they come into power disdain the media that might uncover the truth about corruption once they are in power themselves. The list could go on. Challenges to the rule of law and to civil liberties, lack of recognition of the humanity of one's political rivals, and failure to accept new information that may upset one's ideology but is necessary to meet a societal challenge happen in both East and West. Ellul's call for people to live in such a way that they accept tensions between groups yet recognize a common measure among all is needed more than ever.

Bibliography

Alexander, Michelle. *The New Jim Crow: Mass Incarceration in the Age of Colorblindness.* New York: New Press, 2010.

Asian Human Rights Commission. "THAILAND: Regularization of Torture Following Coup." *Prachatai English*, October, 16 2014. https://prachatai.com/english/node/4409.

Bangkok Pundit. "The Vicious Cycle of Thai Politics Continues UPDATE: Earlier Diagram Located." *Asian Correspondent*, June 6, 2014. https://asiancorrespondent.com/2014/06/vicious-cycle-of-thai-politics/.

BBC News. "Profile: Thailand's New King Vajiralongkorn." December 1, 2016. http://www.bbc.com/news/world-asia-38126928.

———. "Profile: Thaksin Shinawatra." June 24, 2011. http://www.bbc.com/news/world-asia-pacific-13891650.

———. "Thailand Military Seizes Power in Coup." May 22, 2014. http://www.bbc.com/news/world-asia-27517591.

Bhumibol Adulyadej, King of Thailand. "His Majesty the King's Birthday Speech (Thailand)." *Free Republic*, December 4, 2005. http://www.freerepublic.com/focus/f-news/1534028/posts.

Blue Sky Channel TV (Thailand). "BLUE SKY LIVE International." YouTube, February 12, 2014. https://www.youtube.com/watch?v=057JbKJP9h8&t=21s.

---

54. A slogan used in northeast Thailand for Yingluck Shinawatra's 2014 campaign was, "Ready to work with every sector, every region." Yingluck Shinawatra and Pheu Thai Party, campaign posters for 2014 general election.

Buddhadāsa Bhikkhu. *The Buddha's Doctrine of Anattā: A Comparative Study of Self and Not-Self in Buddhism, Hinduism, and Western Philosophy*. Bangkok: Vuddhidamma Fund, 1990.

Campbell, Charlie. "Life in Pictures: King Bhumibol Adulyadej of Thailand." *Time*, October 13, 2016. http://time.com/4526090/thailand-king-bhumibol-adulyadej-life-in-pictures/.

Chai-Anan Samudavanija. *The Thai Young Turks*. Singapore: Institute of Southeast Asian Studies, 1982.

Cochrane, Liam. "Buddhist Faces 15 Years in Jail for Questioning Historical Accuracy of Royal Elephant Battle." *Australian Broadcasting Corporation News*, October 12, 2017. http://www.abc.net.au/news/2017-10-12/buddhist-sulak-sivaraksa--faces-lese-majeste-charge/9040540.

Connors, Michael Kelly. "The King Can Do Wrong." *Sovereign Myth*, September 29, 2008. http://sovereignmyth.blogspot.com/2008/09/king-can-do-wrong.html.

Ellul, Jacques. *Anarchy and Christianity*. Translated by Geoffrey W. Bromiley. Grand Rapids: Eerdmans, 1991.

———. *Apocalypse: The Book of Revelation*. Translated by George W. Schreiner. New York: Seabury, 1977.

———. *The Ethics of Freedom*. Translated by Geoffrey W. Bromiley. Grand Rapids: Eerdmans, 1976.

———. *Money and Power*. Translated by LaVonne Neff. Downers Grove, IL: InterVarsity, 1979.

———. *The New Demons*. Translated by C. Edward Hopkin. New York: Seabury, 1975.

———. *The Political Illusion*. Translated by Konrad Kellen. New York: Knopf, 1967.

———. "Rappels et réflexions sur une théologie de l'État." In *Les chrétiens et l'État*, by Jacques Jullien, Pierre L'Huillier, and Jacques Ellul, 129–80. Paris: Maison Mame, 1967.

———. *The Subversion of Christianity*. Translated by Geoffrey W. Bromiley. Grand Rapids: Eerdmans, 1986.

———. *The Technological Society*. Translated by John Wilkinson. New York: Knopf, 1964.

Fuller, Thomas. "Loved and Hated, Former Premier of Thailand Is Erased from Textbook." *New York Times*, September 16, 2014, A8. http://www.nytimes.com/2014/09/16/world/asia/loved-and-hated-thaksin-shinawatra-former-premier-of-thailand-is-erased-from-textbook.html.

———. "Thai Opposition Party Will Boycott February Elections as Large Protests Continue." *New York Times*, December 22, 2013, A20. http://www.nytimes.com/2013/12/22/world/asia/thai-opposition-party-will-boycott-february-elections-as-large-protests-continue.html?_r=0.

———. "Thai Prime Minister Ordered Removed from Office." *New York Times*, May 8, 2014, A5. http://www.nytimes.com/2014/05/08/world/asia/court-orders-thai-leader-removed-from-office.html?ref=topics.

Goddard, Andrew. *Living the Word, Resisting the World: The Life and Thought of Jacques Ellul*. Carlisle, Cumbria, UK: Paternoster, 2002.

Hawley, Samantha. "Thailand Constitutional Court Rules February General Election Was Invalid." *Australian Broadcasting Corporation News*, March 21, 2014. http://www.abc.net.au/news/2014-03-21/an-thai-court-rules-febuary-2-election-invalid/5337606.

Head, Jonathan. "How King Bhumibol Shaped Modern Thailand." *BBC News*, October 13, 2016. http://www.bbc.com/news/world-asia-33956560.

Human Rights Watch. "Thailand: Junta Entrenched 3 Years After Coup: Empty Promises to Respect Rights, Restore Democracy." *News* (Human Rights Watch), May 21, 2017. https://www.hrw.org/news/2017/05/21/thailand-junta-entrenched-3-years-after-coup.

———. *World Report 2005*. New York: Human Rights Watch, 2005. http://www.hrw.org/legacy/wr2k5/wr2005.pdf.

Jackson, Peter A. "Virtual Divinity: A 21st-Century Discourse of Thai Royal Influence." In *Saying the Unsayable: Monarchy and Democracy in Thailand*, edited by Søren Ivarsson and Lotte Isager, 29–60. Copenhagen: NIAS Press, 2010.

Jory, Patrick. *Thailand's Theory of Monarchy: The Vessantara Jātaka and the Idea of the Perfect Man*. Albany: State University of New York Press, 2016.

Knitter, Paul F. *Without Buddha I Could Not Be a Christian*. Oxford: Oneworld, 2009.

Likhit Dhiravegin. *Thai Politics: Selected Aspects of Development and Change*. Bangkok: Tri-Sciences Publishing, 1985.

MacKinnon, Ian. "Rioting Protesters Close Bangkok Airport." *Guardian*, November 25, 2008. https://www.theguardian.com/world/2008/nov/26/peoples-alliance-for-democracy.

MacKinnon, Ian, and Peter Walker. "First Flight Reaches Bangkok as Airport Blockade Ends." *Guardian*, December 3, 2008. https://www.theguardian.com/world/2008/dec/03/bangkok-flights-resume.

Marshall, Andrew MacGregor. *A Kingdom in Crisis: Thailand's Struggle for Democracy in the Twenty-First Century*. London: Zed Books, 2014.

McIlroy, David. "How Is the Rule of Law a Limit on Power?" *Studies in Christian Ethics* 29 (2016) 34–50.

Merisa Skulsuthavong. "Suthep's Romantic Tale." *New Mandala*, February 14, 2014. http://asiapacific.anu.edu.au/newmandala/2014/02/14/sutheps-romantic-tale.

Pavin Chachavalpongpun. "A Very Wealthy Monarch Grows Wealthier." *Japan Times*, September 9, 2017. https://www.japantimes.co.jp/opinion/2017/09/09/commentary/world-commentary/wealthy-monarch-grows-wealthier/.

Phraezila. "'If They Get Sent to Prison, I Pardon Them'—Said Thai King." YouTube, April 13, 2007. https://www.youtube.com/watch?v=1DZD17stiHI.

Po Saemlamchiak, ed. *Chumnum Nitan Chadok Puenban Khu Mueang Thai*. Bangkok: Amnuaisat, 1982.

Rock, Michael T. *Dictators, Democrats, and Development in Southeast Asia: Implications for the Rest*. New York: Oxford University Press, 2017.

Saxer, Marc. "How Thailand's Middle Class Rage Threatens Democracy." *Social Europe*, January 23, 2014. https://www.socialeurope.eu/2014/01/thailands-middle-class/.

Streckfuss, David. *Truth on Trial in Thailand: Defamation, Treason, and Lèse-Majesté*. London: Routledge, 2011.

Taylor, Adam, and Anup Kaphle. "Thailand's Army Just Announced a Coup: Here Are 11 Other Thai Coups since 1932." *Washington Post*, May 22, 2014. https://www.washingtonpost.com/news/worldviews/wp/2014/05/20/thailands-army-says-this-definitely-isnt-a-coup-heres-11-times-it-definitely-was/.

Vorasak Jandamit, compiler. *Pāli Recitations (with English Versions)*. Bangkok: Mahachula University Press, 1989.

Tan, Hui Yee. "Murder Charges Dismissed Against Former Thai PM Abhisit and Former Deputy Suthep." *Straits Times,* August 28, 2014. http://www.straitstimes.com/asia/se-asia/murder-charges-dismissed-against-former-thai-pm-abhisit-and-former-deputy-suthep.

Yingluck Shinawatra and Pheu Thai Party. Campaign posters for the February 2014 general election. Seen by the author while traveling between Ubon Ratchathani and Roi-Et provinces of northeast Thailand, January 2014.

15

The Hashtag Comes First
Jacques Ellul and Occupy Wall Street

BY Zachary Loeb

> *Zachary Loeb is a graduate student in the History and Sociology of Science at the University of Pennsylvania. Loeb's research focuses on the intersection between the history of technology and the history of ideas, particularly the belief that humanity's romance with technology will lead to disaster. As part of this work, Loeb explores the ideas and legacies of critics of technology such as Lewis Mumford, Jacques Ellul, Günther Anders, and others. Loeb worked as a professional librarian for a number of years, and participated in the 2011 Occupy Wall Street movement as a member of the People's Library working group.*

* * * *

ONE OF THE FIGURES commonly encountered at twenty-first century protests is a character in the most literal sense. That would be the hashtag ("#"). It is a character that is difficult to pronounce, for the proper way to pronounce "#" is on the Internet. The "#" is an aspect of the ways in which new media platform and high-tech devices have been used by activists. These high-tech protests—or "Facebook Revolutions" as they are sometimes called—outmaneuver the control of media gatekeepers by tapping into people's social networks, galvanizing them using videos shot on smartphones that are uploaded to YouTube, while the movement's slogans bounce around Twitter emblazoned with hashtags. Though these devices and platforms may inspire a sort of anarchic glee in the moment it is less clear what the deeper implications of this technological enthusiasm are for these movements.

The work of Jacques Ellul provides a provocative retort to the sort of techno-enthusiasm that has become a fellow-traveler of protest movements in the second decade of the twenty-first century. Indeed, the very notion of "Facebook Revolutions" seems a sort of darkly comic confirmation of Ellul's quip that "we console ourselves in our inability to make a revolution by affixing that word to everything in sight."[1] While Ellul did not live to see city streets filled with activists punching hashtagged messages into their smartphones his work suggests that though he would not have been pleased by such a turn of events, neither would he have found it surprising.

For movement hashtags, protests organized on Facebook, smartphones used for livestreaming, and videos uploaded to YouTube are all demonstrative of the ways in which "technique integrates everything."[2] Ellul had defined technique as the rationally reached "totality of methods" that have as their goal total efficiency in all spheres of activity,[3] he cast it as "a set of rules for the game,"[4] noted that it can "leave nothing untouched,"[5] and highlighted that it is not only that technique "is destroying all other civilizations" but that "it is a whole civilization in itself."[6] It is therefore not that these are movements against the domination of "technical civilization," but that they bear all of the markers of that civilization. The Occupy Wall Street movement provides a prime example of how a protest movement entangled by technique looks.

From the very beginning Occupy Wall Street (OWS) tailored itself as a movement to be digitally disseminated. The initial call to action in *Adbusters* magazine, dubbed the movement "#OCCUPYWALLSTREET." First comes the hashtag, then comes the revolt. The initial call, a work of inspired agitprop, depicted a ballerina balanced atop Wall Street's iconic "Charging Bull" statue against a smoke-filled background from which shadowy figures (some police, some activists) advanced ominously.[7] At the top of the two-page spread words in red provocatively asked "what is our one demand?" while words in white at the bottom of the page read "#OCCUPYWALLSTREET. September 17th. Bring Tent." The presence of the "#" made clear that one initial demand was to spread the movement's name on Twitter.

---

1. Ellul, *Autopsy of Revolution*, 197.
2. Ellul, *Technological Society*, 128.
3. Ellul, *Technological Society*, xxv.
4. Ellul, *Technological Society*, 97.
5. Ellul, *Technological Society*, 125.
6. Ellul, *Technological Society*, 126.
7. Lasn, *Adbusters America*, 48

This chapter will offer a critique of the Occupy Wall Street movement through the lens of Ellul's thinking about technology and political protest. Though I participated in OWS, and its afterlife, as a member of the People's Library working group, this chapter is not intended as a personal reflection.[8] The question of whether OWS succeeded or failed is beyond the ambit of this chapter, though as this chapter will hopefully demonstrate, any attempt to answer that question must critically engage with what technology meant to the movement.

Occupy Wall Street: A Revolt from Within Technological Society

The reason why the initial pages calling people to "occupy" remain worthy of consideration is because people actually descended on Wall Street on September 17th, 2011. And many of them did in fact bring tents. They also brought sleeping bags, food, books, their friends, their hopes, and many of the high-tech gadgets that have become common features of daily life for many people in twenty-first century America.

What began as a small gathering on the morning of September 17th eventually turned into a short march to Zuccotti Park which became the Wall Street location that was "occupied." But, perhaps most significant is that when the sun came up on the morning of September 18th, people were still in the park. At first the occupation was treated as little more than a curio, but armed with smartphones, laptops, and experience with alternative media Occupy activists were able to get the movement's message out without relying on the mass media. Even if in the early days much of what "getting the message out" resembled was simply calling attention to the fact that people were actually occupying Wall Street, and that they were still there.

If much of the media was initially content to ignore OWS, this tenor began to change when footage of officer Tony Bologna pepper-spraying two female protesters trapped behind an orange police net went viral online, even though this was footage shot on a smartphone and uploaded to YouTube, as opposed to something captured by the professional cameras of a major news network. Similarly, when hundreds of protestors were arrested as they attempted to march across the Brooklyn Bridge (on October 1) it made clear that the movement was not going to peter out. The short "Declaration of the Occupation of New York City," which had been accepted by the General Assembly on September 29, ended with a summons to others to "exercise your

---

8. For my personal reflections on the Occupy movement—and a history of the People's Library—see Loeb and Taylor, "Librarian Is My Occupation."

right to peaceably assemble; occupy public space. . . . Join us and make your voices heard."[9] And soon enough versions of "occupy" were appearing in major cities, small towns, and everywhere in between; with some estimating that there were on the order of 600 "occupy" branches in the United States by the time the movement was crushed.[10]

Zuccotti Park, or "Liberty Park" as it was renamed by those occupying it, became something of a carnivalesque city within a city. Against the backdrop of drumming, the park became a place where a person could get a meal, obtain clothes or a sleeping bag, participate in heated political discussions, join any number of working groups, fall in love, browse through the library, or participate in the General Assembly. OWS withstood an attempt on October 14 to clear the park, and managed to weather an unseasonably early snowstorm that hit Manhattan on October 29, and throughout it all the movement continued to grow. Though the movement was actively planning on how to resiliently weather the approaching winter, these plans were to never come to fruition. In the early hours of November 15, Zuccotti Park was raided, the activists who refused to leave the park were thrown in jail, while the tents and the books were thrown into dumpsters. This was part of a coordinated sweep that saw the largest of the encampments similarly, and nearly simultaneously, demolished. And though attempts were made in the weeks and months that followed to re-establish encampments such efforts proved unsuccessful, and the activists who had been brought together by Occupy dispersed to other locales and projects.

In truth, it is difficult to generalize across the various encampments: OWS was different from Occupy Oakland, which was different from Occupy Atlanta, and so forth. Furthermore, the experiences in each place were refracted through the facets of each participant's identity. And though a lot of organizing went into Occupy before that fateful September morning,[11] there's an undeniable level of truth to the organizer Marisa Holmes's comment that "most of us did not come to Liberty Plaza prepared for an occupation. . . . At most we anticipated staying for three or four nights."[12] While many of the obituaries of the movement have been quick, and correct, to highlight Occupy's failings—such as its treatment of race, gender, and colonial histories—one aspect of the movement that has largely received a positive treatment is the way in which the movement made use of social media and readily available computer technology.

9. "Declaration of the Occupation," 148–49.
10. Castells, *Networks of Outrage*, 163.
11. Graeber, *Democracy Project*, 3–54.
12. Holmes, "Center Cannot Hold," 153.

While some activists may scoff that "there are no 'Facebook Revolutions'"[13]—one of the lasting narratives about the movement fits with Manuel Castells's claim that "Occupy Wall Street was born digital. The cry of outrage and the call to occupy came from various blogs . . . was posted on Facebook and spread by Twitter." Castells continues, "Twitter became an essential tool for internal communication in the camps, as well as for linking to other occupations and for planning specific actions."[14] In this framing Castells locates OWS as the United States's manifestation of a wave of protests that swept the globe—including Los Indignados in Spain and the "Arab Spring"—a disparate set of movements about which the common thread seems to have been their canny embrace of new media platforms.[15] Opining on "Why did it work?" David Graeber, who was amongst the early organizers of the Occupy movement, notes that "the omnipresence of phone cameras, Twitter accounts, Facebook, and YouTube ensured" that images and ideas from the camps "could instantly spread to millions" even as he remarks that it is odd that the media "did not play its usual role of presenting only the official police point of view."[16] Granted, his own framing suggests that part of the reason the media could not "play its usual role" was precisely because of "the omnipresence of phone cameras."

Drawing upon the findings of the "Occupy General Survey," Christian Fuchs undermines Castells's claim that "contemporary social movements emerged from and are largely based on the Internet and live and act through digital media."[17] After all, in order for a video of people being pepper-sprayed or news of a bridge being shut down to go "viral," there have to be people offline actually doing those things. And though it may be comforting to those critical of the enthusiasm for new media technologies to note that a high percentage of Occupy activists emphasized the importance of "word of mouth" communication, an extremely close percentage cited "Occupy websites" and an even higher percentage pointed to "Facebook" as sources they relied on.[18]

13. The Invisible Committee, *To Our Friends*, 99.
14. Castells, *Networks of Outrage*, 171.
15. Ratto and Boler, "Introduction."
16. Graeber, *Democracy Project*, 62.
17. Fuchs, *Social Media*, 196–97.
18. Fuchs, *Social Media*, 197. To the question, "These are some sources that you might or might not use for news and information about the Occupy Movement. Please indicate whether you used these sources for news and information about the Occupy Movement," 85.2 percent said "Word of Mouth," 83.4 percent said "Occupy websites," and 89.7 percent said "Facebook." YouTube came in at 72.2 percent, Livestreams at 61.4 percent, and Twitter at only 41.9 percent.

In short, Castells may have been overly enthusiastic, but he was not exactly wrong. Those who conducted the survey emphasized that "word of mouth was the most crucial form of communication" even while noting that "social media was quite important," after all, "respondents . . . arrived at the survey via Facebook."[19] This survey, which was conducted online and only in English, was non-random and solicited the feedback of "several thousand self-identified Occupiers."[20] Based on the "type of participation in the movement" reported by these activists it may be fair to deduce that they were more likely to be in and around the parks than simply following the movement from home. Thus the importance of "word of mouth" may actually be exaggerated by the survey, as these were respondents who were physically in a place where such a method of communication could be useful. True, the survey showed that "contemporary social movements" are not simply "based on the Internet" and they do not exclusively "live and act through digital media"—but the survey still makes clear that "the Internet" and "digital media" were recognized as key tools by Occupy activists.

And in the end it is worth returning to the beginning. While there is much to be said about Occupy, it should not be forgotten that it was a movement that was framed from the outset as being a mediated movement. After all, *Adbusters* did not call it Occupy Wall Street, but #OCCUPYWALLSTREET.

Ellul and the Critique of High Tech "Revolt"

A sentiment such as Jacques Ellul's that technology "destroys the revolutionary impulse by increasing conformity to its own integrated structure"[21] is a serious affront to a movement that encourages its supporters to use a hashtag. And though the retort may be offered that Occupy used these various high-tech devices and platforms as means for advancing the movement's ends, Ellul offers the counter that "ends are incapable of justifying anything because they do not exist: at most they are intentions, ideologies, programs,"[22] or as he put it in another of his texts "the end . . . has been effaced by the means."[23] Ellul's thought offers an unwavering critique of technological society, and by extension of the movements that appear in

---

19. Occupy Research Data Center, "Research By and For the Movement," 69–73. For the purposes of disclosure, it should be noted that the author of this piece participated in the survey of the movement.

20. Occupy Research Data Center, "Research By and For the Movement," 70.

21. Ellul, *Autopsy of Revolution*, 178–79.

22. Ellul, *Critique of the New Commonplaces*, 302–3.

23. Ellul, *Presence of the Kingdom*, 51.

that context. Ellul's thought is an antidote to the techno-optimism that is embedded in terms like "Facebook Revolutions" and a reminder that when a movement puts the "hashtag" first—they may be giving technique the preferential position.

In 1954, Ellul warned of the dangers of a society dominated by technique, but by 1988 he was writing of technique that now "we live incontestably in a society that is totally made by it and for it."[24] Occurring in 2011, the Occupy movement took place "incontestably" within technological society, without actually being a revolt against technological society. Whereas Ellul observed that "technique integrates everything"[25] and that it has "mastered *all* the elements of civilization"[26]—a movement like Occupy reveals the ways in which even acts of resistance become "by" and "for it." The smartphone and social media have given activists and organizers powerful tools for spreading messages and exposing injustice, but these tools serve to entangle these messages and causes within the larger advance and dominance of technique itself. Ellul's critique of technique is not a crude attempt to cut off rebellion at the contradiction, but is instead a rigorous reframing of the question that asks: what is actually being rebelled against?

The confidence in being able to harness technology for rebellious ends speaks to the triumph of the pernicious belief that "the machine is neutral," a commonplace that serves to "humanize" technology and reassure people that they remain in control.[27] Indeed, as Ellul put it, "when we say that we regard technics as neutral, we really think, at bottom, that it is good." Technologies are seen as enhancing human powers and thus "the means are justified by the power which they give."[28] Thus "the means" that enable activists to shoot and disseminate a video of a young woman being pepper sprayed by a police officer or allow them to "live tweet" the shutting down of the Brooklyn Bridge become "justified" in light of "the power which they give." Yet possessing technological power in the moment does not necessarily translate into lasting political power, or actual political change. Ellul provocatively asked "Do we want democracy? Then we should begin by stopping all television and computer games!"[29] But what a movement like Occupy, rather tragi-comically seems to demonstrate, is the way in which "democracy" can become equal parts television

24. Ellul, *Technological Bluff*, 12.
25. Ellul. *Technological Society*, 6.
26. Ellul. *Technological Society*, 127. Italics in original.
27. Ellul, *Critique of the New Commonplaces*, 226–35.
28. Ellul, *Presence of the Kingdom*, 59.
29. Ellul, *Technological Bluff*, 365.

show ("did you see that viral video from the protests?") and video game (who can capture the best picture/video/tweet?).

The bulk of the core technologies whose roles were praised in regards to Occupy appear within the sphere of computer technologies: laptops, smartphones, social media platforms, websites, e-mail, and the like. And the affordances of these gadgets and platforms were among the elements that were seized upon and celebrated by Occupy activists. Decades earlier Ellul was already aware of the hopes that would be hung on the shrinking of computer technologies to the point where they could be widely used. As Ellul warned "the microcomputer is not going to lead to freedom but to conformity within the technical system and to smoother acceptance of the system. It will acclimate us to the computerized world. . . . It is the toy which the technical society provides in order to make more acceptable the concentration that is taking place in every area of the computer system."[30] By allowing devices like smartphones, and platforms like Facebook, to take on the appealing sheen of leading to greater freedom such devices are able to help mollify the rebellious urges that might otherwise be turned against the technical system itself. Allowing activists to believe that they are being granted a share in the computer's power helps to "acclimate" them and to ensure that their revolt will not be against these machines.

Here the particular platforms are less important than the "conformity within the technical system;" the Facebook-using Occupy activist of yore may be the Slack user of today, but what matters is that they have been thoroughly acclimated "to the computerized world." An atmosphere thoroughly favorable to computers develops and thus, "every computer gadget will succeed because it is carried by the mood of our society. The computer is our salvation."[31] Occupy may have aimed to unsettle the "mood" of political complacency, but it did not aim to change this "mood" of technological enthusiasm.

It did not surprise Ellul that in such a social context technology eventually comes to be seen as "the god who saves."[32] The sentiment that "technology is the instrument of liberation"[33] finds its reification in every livestreamed protest and every Tweet featuring the hashtagged name of a movement. The problems that are being protested may be complex; but the simplicity of the technology makes it seem that all that is needed is to land

---

30. Ellul, *Technological Bluff*, 111.
31. Ellul, *Technological Bluff*, 268.
32. Ellul, *New Demons*, 73.
33. Ellul, *New Demons*, 73.

upon the correct technical solution.³⁴ Thus, technology becomes idealized "one refuses to see it for what it is and reduces it to a wonderful working sketch. . . . One passes from the current situation to the situation in which idealized technology will function without any drawbacks."³⁵ And yet the activist's faith in technology's liberating potential, the bureaucrat's faith in technological solutions, and the engineer's desire to bring even more areas of life under technology's aegis—are all symptoms of a society in which technique has become the dominant power.³⁶

Ellul warned against what he termed "vulgarized revolution"—in which rebellious zeal functions not to challenge the society, but in fact provides the opening whereby this spirit is defanged and subsumed.³⁷ One can see this in the success of the slogan "we are the 99%." While this slogan may be based on a certain economic truth,³⁸ it is a line that "suppresses revolt for the sake of a slogan."³⁹ A catchy phrase that sounds wonderful when shouted by a crowd, easily adorns a sign, and can easily fit within Twitter's 140-character limit. Yet even as it speaks to a basic truth it obfuscates more than it clarifies for "propaganda is principally interested in shaping action and behavior, and with little thought."⁴⁰ As a system focused on efficiency,⁴¹ technique is excellent at distributing such ultimately superficial declarations.

Ultimately, a tweet that says, "We are the 99%" or "#OccupyWallStreet" does double duty—it is an advertisement for the movement and for the platform. Corporations cannot buy the kind of advertising that they get when their platforms are used to bring people into the streets. It would have been crass and unseemly had *Adbusters* included Twitter's logo in the initial call to action, but by using the "#" they more or less did exactly that. Thus, activism gradually comes to require that one have the proper devices and access to the right platforms—if the organizing is happening on Facebook, it implies that a person needs to be on Facebook. These platforms and devices obtain added sociocultural importance through their usage by activists, as critiques of these platforms and the companies behind them get deflected by pointing to their usage by activists. Or, as Ellul wittily put it, "One cannot

---

34. Ellul, *Political Illusion*, 64.
35. Ellul, *New Demons*, 115.
36. Ellul, *Political Illusion*, 46.
37. Ellul, *Autopsy of Revolution*, 176.
38. Henwood and the Congressional Budget Office, "It Really Is About the 1%," 15–17.
39. Ellul, *Autopsy of Revolution*, 181.
40. Ellul, *Propaganda*, 278.
41. Ellul, *Technological Society*, xxv.

but marvel at an organization which provides the antidote as it distills the poison"[42]—their use in social movements allows technologies to be presented as "the antidote" even as their usage "distills" technique's "poison." It is not that these techniques are not useful, but precisely that "using techniques always pays off in the short term and then brings disaster."[43]

Thus, in the end, after the police raid the parks and people go home, slogans about the 99% get absorbed into rote political discourse while companies get to see their brands lauded for their role in the movement. Activists may not like terms like Twitter Revolution or Facebook Revolution—but such sayings contain an element of truth: in a technological society, if the revolt is not against the forces of technique, then the victors will not be people on the streets, but the system of technique.

Conclusion

The speed with which Occupy inspired encampments sprang up around the US seems a clear testament to widespread discontent and to a deep desire to rebel felt in many hearts. Yet the movement also provides an opportunity to think seriously about what resistance means in a technological society.

The Occupy movement was not simply a bacchanal drunk on the thrill of new media. Though Castell's comments on the Occupy movement highlighted the importance of the Internet he also noted that "the movement's material form of existence was the *occupation of public space*. A space where the protestors could come together and form a community beyond their differences. A space of conviviality."[44] The technological tale of Occupy is not only about smartphones and social media. While Twitter may have been useful in disseminating messages out of Zuccotti Park, inside the park tents, sleeping bags, brooms, forks, and books were also essential technologies. In casting aspersions at the "world of gadgets," Ellul noted that such things "are indispensable for coping with a society that is more and more impersonal."[45]

And while the Occupy camps were certainly filled with gadgets these "space[s] of conviviality" could also be spots which, however briefly, pushed back against that world that is "more and more impersonal." Ellul had lamented that "people are being plunged into an artificial world which will cause them to lose their sense of reality and to abandon their search for

---

42. Ellul, *Technological Society*, 378.
43. Ellul, *Technological Bluff*, 64.
44. Castells, *Networks of Outrage*, 168. Italics in original.
45. Ellul, *Technological System*, 63.

truth."[46] But perhaps, at its best and most convivial, Occupy exemplifies an attempt to pull people out of the artificial world and to challenge them to resume the search for truth together.

Ellul helps us to remember that, "modern technical instruments have their own weight and by themselves change political structures."[47] Individuals who feel disempowered may feel powerful when they wield new devices, but even if these devices appear as tools of liberation they may wind up acclimating their users to technical civilization. Sufficiently conditioned, users "smile when the word [technology] is spoken, but they fly into a mystic rage when one contests the validity of technology."[48] Ellul provides the theoretical and ethical boldness to "contest" this "validity," demonstrating that if one genuinely desires to question a society than one needs to be willing to criticize its technology as well. Ellul's vision of revolt is a step away from the joyful aplomb of the Occupy encampment, it raises the question that technological society finds most threatening: what are we willing to give up? It may have been enjoyable to be in the carnival-like atmosphere of OWS, but as Ellul notes, "far from being revolutionary . . . festivals only sanctify the technological society, according it the primacy that makes it livable and exciting."[49]

Ellul's challenge flows not from a simplistic disavowal of technology[50] but from a willingness to recognize that an unthinking embrace of technology has helped produce the societies against which these movements are ostensibly rebelling. Even as Ellul emphasized the need for revolution against the technological society he recognized the difficulty of such a project. He acknowledged "rejecting that type of society entails self-denial, a choice, an acceptance of poverty (material as well as spiritual and intellectual), and I am not sure that our young people are prepared to give up their cars, their bathrooms, their phonograph records, or their transistor radios."[51] To that list we might today add "or their Twitter accounts, their Facebook pages, their encryption tools, or their smartphones."

Ellul cautioned that "above all things we must avoid the mistake of thinking that we are free,"[52] and an important step in avoiding that mistake

---

46. Ellul, *Technological Bluff*, 337.
47. Ellul, *Propaganda*, 252–53.
48. Ellul, *New Demons*, 73.
49. Ellul, *Autopsy of Revolution*, 287.
50. Ellul, *Autopsy of Revolution*, 275.
51. Ellul, *Autopsy of Revolution*, 274.
52. Ellul, *Technological Bluff*, 411.

is to recognize that the things we think of as the keys to our freedom may actually be our shackles.

Bibliography

"Declaration of the Occupation of New York City." In *We Are Many: Reflections on Movement Strategy from Occupation to Liberation*, edited by Kate Khatib, Margaret Killjoy, and Mike McGuire, 148–51. Oakland: AK Press, 2012.

Castells, Manuel. *Networks of Outrage and Hope: Social Movements in the Internet Age*. Cambridge: Polity, 2012.

Ellul, Jacques. *Autopsy of Revolution*. Translated by Patricia Wolf. Eugene, OR: Wipf and Stock, 2008.

———. *A Critique of the New Commonplaces*. Translated by Helen Weaver. Eugene, OR: Wipf and Stock, 1994.

———. *The New Demons*. Translated by C. Edward Hopkin. New York: Seabury, 1975.

———. *The Presence of the Kingdom*. Translated by Olive Wyon. Colorado Springs: Helmers and Howard, 1989.

———. *Propaganda*. Translated by Konrad Kellen & Jean Lerner. New York: Vintage, 1965.

———. *The Technological Bluff*. Translated by Geoffrey Bromiley. Grand Rapids: Eerdmans, 1990.

———. *The Technological Society*. Translated by John Wilkinson. New York: Vintage, 1964.

Fuchs, Christian. *Social Media: A Critical Introduction*. Los Angeles: Sage, 2014.

Henwood, Doug, and the Congressional Budget Office. "It Really Is About the 1%." In *Occupy: Scenes from Occupied America*, edited by Astra Taylor et al., 15–17. London: Verso Books, 2011.

Graeber, David. *The Democracy Project: A History, a Crisis, a Movement*. New York: Spiegel and Grau, 2013.

Holmes, Marisa. "The Center Cannot Hold: A Revolution in Process." In *We Are Many: Reflections on Movement Strategy from Occupation to Liberation*. Oakland: AK Press, 2012.

The Invisible Committee. *To Our Friends*. South Pasadena: Semiotext(e), 2015.

Lasn, Kalle, ed. *Adbusters America: Journal of the Mental Environment* #97, 19 (September/October 2011).

Loeb, Zachary, and Jaime Taylor. "Librarian Is My Occupation: The Story of the People's Library of Occupy Wall Street." In *Informed Agitation: Library and Information Skills in Social Justice Movements and Beyond*. Sacramento: Library Juice Press, 2014.

Occupy Research Data Center. "Research By and For the Movement." In *We Are Many: Reflections on Movement Strategy from Occupation to Liberation*. Oakland: AK Press, 2012.

Ratto, Matt, and Megan Boler. "Introduction." In *DIY Citizenship: Critical Making and Social Media*, edited by Matt Ratto and Megan Boler, 1–22. Cambridge: MIT Press, 2014.

16

The Illusion of Change

From Greece to Italy and Spain

BY Carmen Madorrán

> *Carmen Madorrán Ayerra holds a PhD in Philosophy with a dissertation on "Human Needs and Ecological Limits in Economic Democracy. A review of David Schweickart's proposal," under the supervision of Jorge Riechmann. She earned her master's degree in philosophical critique and argumentation at the Universidad Autónoma de Madrid (Spain). She is currently part of the teaching and research staff in training in that university, where she focuses on moral and political philosophy. She has published papers on the socio-ecological crisis, basic human needs, consumerism, and collective consumption alternatives. Among her interests are also the problems of historical memory and the repression under Franco's dictatorship.*

* * * *

THE FIRST THING WE have to address in this essay is what we talk about when we speak of change. *Who* is talking about change, and *where*. In the European political context, change acquired a new dimension and meaning during the economic crisis which began in mid-2007, particularly after May 2011 with the surge of the *indignados* movements in Spain (echoed in different countries—like *occupy* in the USA). Underlying these social movements was a rejection of the establishment understood in a broad sense. A fundamental claim and slogan of those days was that the political institutions and representatives *do not actually represent the people*. However, they are not the subject of this paper, since it was not they who introduced the idea of change in the political arena. My focus is on the new parties claiming to embody change through so-called "new" politics. The most important ones have risen—not by chance—in southern Europe. It should be born in mind that the financial crisis hit those southern countries particularly hard. Let us look, briefly, at these parties: their origins and background. In order of appearance, we will talk about Syriza, Five Star Movement, and Podemos.

Syriza is a left-wing coalition of political parties founded in 2004 in Greece. However, until 2007 the coalition didn't receive much attention because of its moderate performance. With the crisis that has reached deep into Greece, Syriza started to increase in both popularity and votes. Now it is the largest party in the Hellenic Parliament, with party chairman Alexis Tsipras serving as Prime Minister of Greece. As we all know, the former government of Greece (PASOK and New Democracy) signed a memorandum accepting the conditions of the "Troika"—the International Monetary Fund, European Commission, and European Central Bank.

The effects of the crises and the measures adopted by the government to achieve their commitments to Europe under that memorandum impoverished the Greek population. They were desperate. They also felt their government had abandoned them (and in my opinion, it had). The rise of Syriza is understandable under those circumstances. Syriza called for a big change in the way politics was understood and performed. They said they wanted the Greeks to take their future in their own hands. Allegedly, they embodied a left-wing approach to the debt crisis and how to deal with it. And overall they insisted on recovering their sovereignty and rejected the austerity policies imposed by the Troika.

Our second example is Italy's Five Star Movement, founded in 2010 by a talented comedian (Beppe Grillo) and an editor and political activist (Gianroberto Casaleggio). Despite its name, Five Star Movement is a political party. It is self-defined as an anti-establishment, ecological, Eurosceptic party. In fact, in the European Parliament they decided to join with the Europe of Freedom and Direct Democracy group, along with the right-wing populist United Kingdom Independence Party. Another of its main characteristics is that they portray themselves as post-ideological and therefore beyond both the left-right cleavage. They usually refer to the Internet as a solution for different political and social problems—not far from cyber-utopianism—and they debate and elaborate some of their positions online. Since 2010 they have gone through ups and downs in Italian political life. However, the local elections held in June 2016 represented a big step forward for Five Star Movement. They won mayoral elections in both Rome and Turin (both electing women in their thirties).

The third example is Podemos, a political party founded in Spain in 2014, with the manifesto *Mover ficha: convertir la indignación en cambio politico* (*Making a move: turning indignation into political change*). The idea stemmed, the previous year, from Izquierda Anticapitalista ("Anticapitalist Left"), a left-wing party founded in 2009, which had never achieved institutional representation. They provided the organizational base and Pablo Iglesias his image. He was unknown to most people in Spain but famous within left groups and social movements because of his local TV show. He had also a certain political prestige because he had been involved in different political movements and parties, and taught political science at the university. Their main positions at the time were against the *casta* (corrupt elite) and against the austerity program. Podemos drew a lot of attention from journalists and mass media because of its alleged connection with *indignado* movement, and they received lots of invitations to present their ideas and leaders on television. They broke into the Spanish political life obtaining almost eight per cent of the vote in the European Parliament election. They have increased their votes and representatives in Spanish institutions in an unprecedented way. This is a party created in 2014 that just two years later won twenty per cent of the vote.

It is worth noting the common features of these three political parties. First of all, they all emerge from similar backgrounds, rooted in and responding to the same problems and symptoms. They each appeared in the midst of a crisis which was not only economic, but more fundamentally a general legitimacy crisis of the European system and institutions. Broadly, the crisis is twofold: economic and political. One cause is that the middle-class golden dream has been shattered. Until now, many citizens were reasonably satisfied with the economic system, enjoying the welfare state and the consumer society. Now, suddenly, people lost their jobs or suffered wage cuts. Furthermore, a set of cutbacks were implemented to meet the debt reduction targets, making some welfare and social services more expensive, especially public health and education. Pensions are also in jeopardy. As a result, a large part of the population (which does not identify with the working-class or with the whole idea of classes, for that matter) is profoundly disappointed in the economic system.

This is compounded by a crisis of confidence in public institutions—political parties, national governments, as well as European institutions—which are blamed for the people's suffering. In Italy, Greece and Spain, political corruption feeds such disaffection.[1] For example, in Spain 485

1. In fact, in recent years, the main concerns of the Spanish population, as recorded by the annual survey conducted by the Spanish Sociological Research Center (Centro de Investigaciones Sociológicas, CIS) are: unemployment, corruption and fraud, and

members of the Popular Party, the recently reelected governing party, are currently charged with corruption, including the former Minister of Economy and the party treasurer. The unacceptable normalcy repeated corruption scandals were handled became one of the central issues of these new parties' speeches. In contrast to the old politics, dark and corrupt, they presented themselves as true representatives of the people. If there is something these three parties have in common, that is their self-perception as the embodiment of change, a necessary change, a change towards a new way, although the content and meaning of this change is mutable or even empty. That hardly seems to matter because the people are fed up and upset. Politicians stink and economic crisis has come to stay!

Political Illusion?

In his book *The Political Illusion* (1967), Jacques Ellul offered a deep analysis of the relationship between state, politics and democracy in his time. It would be impossible to address all the ideas developed in that work. However, some key concepts put forward in *The Political Illusion* may be very useful to analyze from a critical standpoint the phenomenon of so-called "new politics."

In Ellul's words, *political illusion* could be defined as the belief "that the citizen, through political channels, can master or control or change the state."[2] Ellul distinguished three core elements within the political illusion: the *modern state*, with its inherently enormous bureaucratic machinery; *political solutions*; and *political participation*. I will focus on the latter, and its (allegedly) illusory nature, since it is the most insightful part of Ellul's analysis to help examine the political processes currently (and for some years) taking place in certain European countries.

In this regard, three interrelated ideas seem essential:

First, the leading role of "public opinion." In a democracy, the consent of public opinion is required as a legitimating condition for political action. As Ellul puts it, once the masses have entered political life and have a vehicle for expression (i.e. public opinion), they can no longer be pushed out of political life. It is no longer possible to govern against public opinion. Since the masses are now *concerned* about politics, they need to be *informed* (public opinion must be about something, it must have an object). Furthermore, in a globalized world, any and every issue is of interest to the public. People care not only about their local environment or community since we now live in

---

politicians and political parties. Detailed data can be found in CIS, *Estudio nº 3146*, 4.

2. Ellul, *Political Illusion*, 160.

a *global village*. Intermediaries are needed to provide what will necessarily be a simplified account of the world events. Facts are not merely conveyed through what we call *information*. They are mediated—deformed, edited, transformed—into images. As Ellul puts it, "the pure fact has no power at all. It must be elaborated with symbols before it can emerge and be recognized as public opinion."[3] Hence, the totality of our contemporary world is lived only indirectly, as a "retranslated, edited universe,"[4] without any direct relation to any fact. This mediated reality is relatively independent.

Second, political facts are only those matters that receive the attention of public opinion. Since the colossal amount of (edited) information we receive is unmanageable (due to its vastness, complexity, and the prerequisite knowledge essential in so many fields), attention can only be drawn to a portion of it. The internet could be an illustrative example in that regard. It seems to be filled with endless information and facts. However, to gain political relevance, public opinion must be focused. Furthermore, if we expect any political action to be taken, such focus or attention span must have some duration (who remembers yesterday's "trending topic" on twitter? Or the last campaign?).

In this sense, according to Ellul, "it is no longer the fact itself, but the fact translated for public consumption which is now called a political fact."[5] The rest simply does not exist. An archetypal example of a "non-fact" was, for a long time, concentration camps in Europe. There was information available on the subject, a lot of people had experienced them, and yet they were not a matter of public opinion for at least ten years after the war. In Spain, for example, a large majority of the population is still unaware of the existence of concentration and forced labor camps in their own country between 1936–1947. As we have seen, the public has only an indirect knowledge of reality, since it has access only to appearances. And only some of those appearances, through public opinion, become political facts and, therefore, potential political problems. To conclude with Ellul's words, "[t]here is practically no problem by itself"[6] and "those lacking the means of influencing the presentation of facts do not exist in the world of politics."[7]

Third, also following Ellul: "Another aspect of the political illusion is to believe that the citizen *can* effectively participate in political life."[8] As already

3. Ellul, *Political Illusion*, 104.
4. Ellul, *Political Illusion*, 113.
5. Ellul, *Political Illusion*, 100.
6. Ellul, *Political Illusion*, 118.
7. Ellul, *Political Illusion*, 128.
8. Ellul, *Political Illusion*, 163.

mentioned, in order to participate, citizens must be informed (with all the related problems). But, apart from that, what triggers political participation? An important element is people's need to belong, and the search for a meaning beyond the merely individual concerns. In our contemporary societies, one can experience "the communion that he absolutely needs, but no longer finds, in his family, his neighborhood, or his work a common objective, some great popular drive in which he can participate, a camaraderie, a special vocabulary, an explanation of the world. Politics offers him these joys and symbols, these indispensable expressions of communion."[9]

Thus, political participation is presented by Ellul in a sort of religious light. Political participation is, to some extent, a substitute for faith, a cult. At the same time, even if every citizen wants to have a say on every issue, even if he has an opinion about all imaginable subjects, demanding to be heard, claiming for actions to be taken, he himself refuses to be "an active and constructive participant in the power structure or to consider himself responsible for political actions and events."[10]

As one might have expected, such an impulse or need to be part of something finds an outlet in political parties. Ellul considers political parties in democratic regimes only as groups seeking ever greater political power for their own members. In this sense, political activists are compared to soldiers in their equal surrender as individuals to merge into the group. Commitment to a political party is thus rooted in a renunciation of critical thinking as an individual, of having one's own opinion, of assuming personal responsibility. Instead, the party offers security and clear ready-made judgments.[11] As we will see below, the figure of a "leader" is also essential, easing the way to that surrender.

The Illusion of Change

Ellul's core ideas about the political illusion are helpful tools for the analysis of the three movements described above. First of all, these parties all assume the needs for a broad popular support and to command public opinion. They are experts in modern political communication (as already mentioned, they appear in mass media and have an active participation in social networks). They are comfortable in our world of images, thoroughly engaging in the editing tasks we mentioned earlier. All these new parties are fully aware of the importance of communication and mass media.

9. Ellul, *Political Illusion*, 22.
10. Ellul, *Political Illusion*, 166.
11. Ellul, *Political Illusion*, 171.

Significantly enough, the leaders of Podemos (especially its front man Pablo Iglesias) endeavored to host TV shows, appearing as often as possible in political *tertulias* and debate programs. They insistently put forward a repetitive message with a few catchy (arguably void) slogans tapping on the people's frustration, particularly the youth: "we can," "regeneration," "out with the *casta.*" In their own words, they managed to determine the vocabulary and rhetoric of the political chessboard.

Connected with this, the whole political scenario, built around parties and what they present and debate as political problems, has been fostered by a sports-like framing of politics, in turn, favored by show-business and mass media. New parties use symbols better than anyone else. For example: their members always appear without tie or suit highlighting their youth and alternative manners. In Spain's 2016 general election, Podemos presented its political program imitating IKEA's catalogue, showing the houses of their candidates along with their political proposals. Through this brilliant (yet simple) strategy, politics became associated with something familiar to the majority of the population. They constantly appeal to emotions. As Martha Nussbaum puts it, public emotions are generated not only in the rhetoric of political leaders: "[L]eaders lead in many ways. They lead with their bodies, their clothing, their gestures. And government . . . generates public emotion through many strategies: through public artworks, monuments, and parks, through the construction of festivals and celebrations, through songs, symbols, official films and photographs, through the public use of humor and comedy, even by shaping the public role of sports."[12]

Second, as we have seen, political facts and political problems are only those that gather the attention of public opinion. Aware of the people's inability to have a sound opinion on an unmanageable array of subjects, they appeal both to their grassroots and a wider spectrum of voters with simple, emotionally engaging rhetoric. Instead of addressing the variety of traditional left-wing concerns (class struggle, wages, exploitation, finance system, equality, wealth distribution, etc.), they focus on several specific topics to rally together a mass of supporters against a well-defined easily-identifiable enemy. At the early stages, the message is "regeneration," "anti-austerity," "new politics," and the enemy is the Troika, the *casta*, and corrupt political leaders. Public opinion is thus oriented and focused. Attention is not dispersed on multiple things.

Both Podemos and the Five Star Movement define their own strategies as populist, in the sense developed by Laclau and Mouffe, and disregard the

---

12. Nussbaum, *Political Emotions*, 203.

left-right split.[13] Initially, for instance, Podemos and Syriza spoke of a radical change identifiable with the demands of the European left: nationalization of strategic sectors, creation of a public banking, increase of the tax burden, and launch of social protection programs. In a few months, statements became quite different. Not only the tone was softened, but the most radical proposals were withdrawn altogether. Podemos stood for the last general elections claiming that their proposals were social democratic and they stopped talking about *casta*, so there was nothing to fear.

The other aspect criticized by Ellul as an illusion was the idea of citizens effectively participating in political life. That idea (that citizens can actually change things through political participation) is precisely one of the cornerstones of these three parties, and there lies part of their success. For many of their voters it is easy to feel part of the change just by watching their leaders on TV debates (or posting a tweet). As Ellul said, people want to be part of something bigger than their lives, but with the least possible commitment. It could be questioned whether most Podemos's voters actually want a real change that could make their lives worse and less comfortable. They surely do not want to feel part of the problem, part of the *casta* or the corruption. They want to see themselves as something new, young, and fresh, with new ideas and new ways of doing things. So these parties offer that kind of easy identification: for example, the generational issue is introduced in the debate. Youth becomes a value *per se* and synonymous with change, a change which is intentionally left undefined or whose content gradually mutates. They proclaim the new against the old, and they all share what we could call an *Adam complex*, as if only the new existed and anything old was completely useless.

They also oppose professional politicians, on the one hand, and *ordinary* people, on the other. And they identify with the latter, even if they start to attend the same meetings and lead a life similar to those politicians. At the same time, they are allegedly only a transmission channel, the mere political expression of the *ordinary* people that gathered on the streets in the *indignado* movements. They simply convey the message of a frustrated population, they are not political elite; they are not even politicians (because politicians "don't represent us"). Yet, they claim to represent 99% of the people. Ordinary people; movement; change. These are the new mantras. And they claim to be able to bring change. How? By running in elections and being elected to public offices (maybe not all that new, after all). They offer easy explanations and solutions to complex problems, and they know they

---

13. On this issue, see Laclau and Mouffe, *Hegemony and Socialist Strategy*; and Laclau, *On Populist Reason*.

are doing it. They assume that most people cannot devote the time required to understand the complexity of today's problems, and they also assume that some people are unable to understand anything but soccer. Here lies a significant paradox: these parties (particularly in the case of Podemos) are formed by intellectuals who present themselves as common men and women. Žižek's remark on populism is quite illustrative in this regard:

> We all know the old joke about a guy looking for his lost key under the street light; when asked where he lost it, he admits that it was in a dark corner. So why is he looking for it here, under the light? Because the visibility is much better here. There is always something of this trick in populism. It looks for the causes of troubles in the Jews since they are more visible than complex social processes.[14]

In conclusion, Ellul said there were only two options available: to "continue to believe that the road to solving our problems is the traditional road of politics"[15] along with reforms, ideas, left-wing vs. right-wing debates, etc. (all empty shadows to Ellul);

> or we turn away from the illusory debate . . . [and] try to create positions in which we reject and struggle with the state, not in order to modify some element of the regime or force it to make some decision, but, much more fundamentally, in order to permit the emergence of social, political, intellectual, or artistic bodies, associations, interest groups, or economic or Christian groups totally independent of the state, yet capable of opposing it, able to reject its pressures as well as its controls, and even its gifts.[16]

Ellul's proposal to get out of political illusion could be described as "creating such poles confronting the state . . . returning an autonomous vitality to certain parts of society, would make possible a political life that would be something else than mere illusion."[17] Ellul was aware of the potential accusation of utopianism: "I have never said that it is possible. I have only indicated what I consider to be the basic condition for social and political life and the only way to escape the political illusion."[18]

However, Ellul was not proposing some kind of *apoliticism*. He considered that "depoliticization is no remedy at all for the shortcomings of our

14. Žižek, *Žižek's Jokes*, 56.
15. Ellul, *Political Illusion*, 221.
16. Ellul, *Political Illusion*, 222.
17. Ellul, *Political Illusion*, 223.
18. Ellul, *Political Illusion*, 223.

political world, or the increase in state power. . . . The idea that one can escape politics by being nonpolitical is just as absurd as the political illusion itself."[19] On the contrary, he argued that two conditions were required to recover democratic behavior in our societies: it would be necessary to free political matters from its myths and to look at them from a proper perspective. In addition, the personal development of citizens must change in order to enable them to know and understand things independently.

According to Ellul, the new parties proclaiming change work in the world of illusion and participate in the illusion that every problem has a political solution, that the state must be reinforced in order to solve such problems. From his point of view, political parties contribute to spread the idea that politics is what the state does. Hence, the scope of the debate is accordingly constrained within those boundaries. The whole idea of change—which they claim to embody through symbols and propaganda in an attempt to convince people that that is the way to make things better—would then be another illusion. It is only appropriate that the Spanish term "ilusión" has two meanings: excitement or enthusiasm, on the one hand; and, on the other, the English meaning of "illusion," i.e., something that looks or seems different from what it is. Certainly the idea of a political change underlying these "new" parties has a little of both. We will see which one prevails.

Bibliography

Centro de Investigaciones Sociológicas (CIS). *Estudio nº 3146. Barómetro de julio de 2016*. www.cis.es/cis/export/sites/default/Archivos/Marginales/3140_3159/3146/Es3146mar.pdf.

Ellul, Jacques. *The Political Illusion*. Translated by Konrad Kellen. Eugene, OR: Wipf and Stock, 2009.

Laclau, E. *On populist reason*. New York: Verso, 2005.

Laclau, E., and C. Mouffe. *Hegemony and Socialist Strategy*. New York: Verso, 1985.

Nussbaum, Martha. *Political Emotions*. Cambridge: Harvard University Press, 2013.

Žižek, S. *Žižek's Jokes: (Did You Hear the One About Hegel and Negation?)*. Cambridge: MIT Press, 2014.

19. Ellul, *Political Illusion*, 200–201.

17

Ellul and Health Care Reform

by Raymond Downing

> *Raymond Downing (MD, New York) has spent about one-third of his professional career as a medical doctor in the USA and two-thirds in several countries in Africa, currently in the Department of Family Medicine at Moi University in Eldoret, Kenya. He is the author of* As They See It: The Development of the African AIDS Discourse *(Adonis & Abbey, 2005),* Suffering & Healing in America: An American Doctor's View from Outside *(Radcliffe, 2006),* Death & Life in America: Biblical Healing & Biomedicine *(Herald Press, 2008), and* Biohealth: Beyond Medicalization: Imposing Health *(Pickwick, 2011).*

* * * *

EMBEDDED IN THE LONGSTANDING USA debate about healthcare reform is a political question: is healthcare the business of the government? To those on the political left, healthcare must be the business of government, because only government could ensure equitable distribution of basic healthcare to all citizens irrespective of their ability to pay. Those on the right feel basic healthcare is not the business of government; it should remain a matter of choice, provided by the free enterprise system—which, in theory, will hold costs down by competition, and therefore, in theory, be available to all who need it. For both sides, the political process is critical: for the left, to ensure that the state does control and finance healthcare; for the right, to ensure that it does not.

However, absent from both sides of the discussion is the nature of healthcare itself; the debate remains focused on financing. The content of this healthcare "package" is not being fundamentally challenged. Both sides have deferred to the experts and the technicians who rely on "evidence" to decide what must be included. The political question is not about the nature of healthcare, but only about its financing.

Jacques Ellul, in the first chapter of *The Political Illusion*, reminds us that "The only domain in which politics can still act is . . . the sphere of the ephemeral and the fluctuating." He compares this with the sphere of the "necessary"—areas of society designed by and dependent on technicians, and not subject to change through the political system. Decisions in these areas are "made on technological grounds" in which "efficiency [is] the primary aim." It seems obvious that biomedical healthcare is in the realm of the "necessary"; only its financing is "ephemeral" and subject to political manipulation.[1]

Financing is not a trivial question; most European countries make some healthcare available to most of their citizens. The more important question, however, is the nature of healthcare itself, a matter not subject to political change. This chapter will review some of Ellul's early thinking on the nature of medicine, and then consider how this might influence our consideration of healthcare reform in light of *The Political Illusion*.

The Nature of Humans

To understand medicine, Jacques Ellul felt he first needed to establish the nature of human creatures and the relationship of those creatures have with their Creator. This was the topic of one of his earliest essays, "Biblical Positions on Medicine," first published in 1947, and finally available this year in English.[2]

His thesis was straightforward, and at core neither surprising nor unique. People, he said, have "two parts, soul-body and spirit, [which are] closely linked, completely intertwined, to such an extent that no man can differentiate them and separate what is natural in man from what is supernatural." But more than just this link, "the corporeal appears only as a sign of what the spiritual is. . . . The real drama, the real action, takes place on a stage to which we don't have access, where we do not feel comfortable." That corporeal sign is often an illness for which we seek medical help, but biomedical doctors usually don't have a ticket for the spiritual theater, the ultimate source of the illness. They therefore focus on the physical manifestation, which Ellul calls "only a repercussion, only a secondary phenomenon"—only a symptom.

---

1. Ellul, *Political Illusion*, 25–67.
2. Ellul, "Biblical Positions on Medicine." Unless a different source is given, all Ellul quotes are from this essay.

I have considerably condensed his argument. In granting a connection between the spiritual realm—including sin—and illness, he takes pains to point out that "this link between sickness and sin must not be understood in a simplistic sense," such as "he who is the greatest sinner is the one who is most sick—or that sickness is the sign of a greater sin. . . . Not at all." However, "the healing of sickness without the forgiveness of sins is only an adjournment, a patching up, a little boost: it is not health. This deliverance from sickness has no value in itself. It can [only] be a temporary betterment."

"Illness," he says, "possesses a profound meaning . . . and the doctor obviously should be careful not to deflect sickness from its meaning." Unfortunately, biomedicine cannot tell us what that meaning is. Thirty years after Ellul's essay, Susan Sontag wrote a polemic against the cultural meanings of illness she saw—meanings still present, perhaps, because of the remnant of understanding in our culture that illness *does* have meaning. In her writing, however, she wanted "not to confer meaning . . . but to deprive something of meaning." She was troubled by the inappropriate and damaging metaphors of illness she confronted, and wrote to demonstrate "that illness is *not* a metaphor."[3] Ironically, she was left with only biomedicine, and betrayed a confidence and faith in it far beyond the confidence I as a doctor hold.

The Relationship of Humans with Their Creator

The human person, Ellul writes, "is not an autonomous being," but rather "wholly dependent on the Creator." And again: "between the creature and the Creator there can be a right order of relation, and this is what can promote health best." Later, he expands a bit: "Left to ourselves, we will go immediately to death by the way of sickness. It is God's hand that restores us continually in a state of relative health, which we do not deserve. Healing is thus nothing other, in every situation, than God's merciful intervention in the course of nature."

Ivan Illich developed this concept in his continuing analysis of medicine in his writings after his well-known *Medical Nemesis*. Referring to the medieval view that all of nature was alive, Illich says that between the fourth and the fourteenth centuries, people believed that the "birthing power of nature was rooted in the world's being contingent on the incessant creative will of God."[4] (In the words of the old spiritual, "He's got the whole world in his hands.") However, when people began no longer to believe in the sustaining care of God, they developed tools to sustain

3. Sontag, *Illness as Metaphor*, 3, 93, 102.
4. Illich, "Brave New Biocracy," 4.

the life and health they had previously believed was God's realm.[5] As we will see, Ellul calls this use of tools idolatry: "Sometimes medicine is even presented to us as completely contrary, opposed to God's will, a sign of man's revolt against God (2 Chr 16:12, Jer 17:5). . . . It truly takes the place of God and is for this very reason condemned."

Physical and Spiritual Healing

Nevertheless, Ellul affirms attention to both the physical and the spiritual, but it can be difficult to get the balance right. We may be tempted to view a patient's spiritual condition mechanistically. As the patient incrementally repairs the creature-Creator relationship, we hope for a corresponding improvement in the physical symptoms. Here we are viewing our physical and spiritual lives as being linked like gears, with movement in one causing immediate movement in the other.

Or, we can try to treat each part separately, essentially de-linking the gears. We rely on the motto "We treat, God heals," assuming that our treatment will somehow catalyze God's healing intervention. And while doctors may tell some stories that illustrate this connection, there are plenty of stories that show the opposite, the first of which is Job's. We, like Job and his friends, have trouble getting it right.

Ellul, far from resolving this dilemma, simply affirms it. "Thus the healing of sin affects the cause of the illness, which is always a disobedience to this natural order that God established. Of course, the symptoms of the disease, its material consequences, are not ended thereby. . . . Thus the Christian notion of sickness indeed entails a material healing and activity as well."

As with most of Ellul's writings, there is no agenda here, no program to follow, no principles that translate easily to the construction of a "Christian healthcare system." He does not eschew the doctor's work, but tells us only that "treatment is given to the doctor by God, that it is indeed an acceptable means of caring for the body." His task is not to eliminate medicine, but to help us see it in perspective. But note: his focus is on addressing the meaning of illness, healing, and relationship with the creator, without respect to medical technology. His analysis applies today as well as in Biblical times. The presence of advanced medical technology does not supplant or obviate the importance of this analysis; on the contrary it highlights it.

---

5. Illich and Cayley, *Rivers North of the Future*, 64–79.

Counterproductivity

The difficulty we have with meanings of illness, mentioned above, and the temptation to deny them altogether, is where Ellul's 1947 argument begins to anticipate so much of what he later wrote about technology. He suggests that biomedical treatment is not only incomplete, but could also be counterproductive. Denying meaning that is there is certainly counterproductive, because it leads us away from healing. There is a similar dynamic when biomedicine (successful productive biomedicine) "raises hope and stimulates faith." In doing so "medicine dresses itself up in what is not its own. It draws forth the praise and gratitude that are due only to God." This is "when medicine becomes an idol, a power that we petition independently of God." Any idol, whether secular or spiritual, is counterproductive precisely because it is false.

But there are other more direct forms of counter-productivity that Ellul mentions. For example: "We observe that man succeeds somewhat in removing suffering, but not in overcoming or reversing sickness. For if a sickness recedes, how many other forms reappear or arise for the first time?" The question was speculative, but half a century later research seemed to show that Ellul was on the right track. In the last decade of the twentieth century there was a study of treatment methods for newly diagnosed early prostate cancers: half received surgery, and the other half didn't. Those with surgery were less likely to die of prostate cancer, but six years after diagnosis overall death rates in both groups were the same. In other words, "prostatectomy does not change the date of death; all it changes is the likelihood that prostate cancer will be the direct cause."[6]

Ellul goes on: "If acute illness is arrested, how much more does our resistance and general health weaken? If microbial diseases seem to be conquered, how much more do nervous diseases arise, and so forth?" Again, recent research confirms Ellul's insight. Considering cancer survivors, those people with a diagnosis of cancer who have been treated and are still living, studies in the last three decades have shown the following: "Compared with their peers, cancer survivors experience significantly decreased quality of health; increased incidence of chronic health conditions; increased levels of psychological disability; and other physical, emotional, and financial challenges."[7] We may have defeated the cancer, but we clearly did not defeat ill health.

---

6. Hadler, *Last Well Person*, 96.
7. See Sunga et al., *Care of Cancer Survivors*.

And finally, Ellul says, because of our individualistic and materialistic approach to remedies, we are left with "only one goal: to end suffering." In doing so, "we have lost the sense of life's relativity and of the individual person's integration within real communities and generations. All of this falsifies the idea of treatment. The true cure is the one that attains the roots of the illness and that acts over a more or less extended period of time, that may even act only in our descendants. "To twenty-first century ears, this sounds like gene therapy, but gene therapy does nothing to situate us in our communities and with our ancestors and descendants." Symptom relief remedies, which do not "reach illness in its roots," are ultimately counter-productive because they draw attention away from the true nature of the illness. True healing, as Ayi Kwei Armah demonstrates in his novel *The Healers*, is healing not just of disease, but of entire communities.[8]

Healthcare Reform

Until now, we have been discussing medicine historically. We have not yet considered healthcare in all its technological glory, as a key part of the technological society. Reforming this technological medicine is a political matter that we must now confront.

Healthcare reform, especially in the USA, is a perpetual concern. Since healthcare is expensive and not available to everyone, we keep seeking ways to expand the coverage and lower the cost at the same time. Yet we haven't really questioned the nature or content of this healthcare system which we want to make more widely available. We assume everything medicine offers for illness is needed and is helpful. If a scientific study proves medical or surgical intervention for a certain problem lengthens life, or reduces pain, or helps us sleep better, or improves our sex life, then it should be offered to everyone who has that problem. We do not discuss whether we *ought* to continue having regular sex after our bodies have naturally slowed down because of age. We do not seriously debate whether organ transplantation or replacement *ought* to be performed—for anyone.

We, doctors as well as patients, are encouraged to eliminate *all* pain. There is a vicious cycle of doctors giving patients marginally helpful treatments because they think patients expect *some*thing, and patients taking those treatments (even though they may not help) precisely because the more knowledgeable doctor prescribed them. We have become convinced that the most important prevention is offered by the medical system:

---

8. Armah, *Healers*.

screening tests and preventive *medicines*—forgetting that how we move and eat are far more important in keeping us healthy.

In other words, we do not discuss healthcare reform in the US; we discuss healthcare *financing* reform: how can we finance the system we have? But as we saw at the beginning of this chapter, financing is an ephemeral issue. The more important issue, the nature of that system, is "necessary" and not really subject to political change. So we get derailed: instead of looking honestly at that system to see what it really accomplishes, we concentrate mostly on making it more efficient.

Ellul would not be surprised. Efficiency was for Ellul a pivotal characteristic of technology. His definition of the essence of technology—he called it "technique"—was "the *totality of methods rationally arrived at and having absolute efficiency* (for a given stage of development) in *every* field of human activity."[9] When we realize that all methods that people develop to accomplish tasks efficiently can be considered "technology," it is clear that healthcare is right at the center of our technological society.

While efficiency is the defining characteristic of technological systems, it is not the only characteristic Ellul describes. Consider briefly these others as they apply to healthcare:[10]

Technological systems are automatic and self-augmenting: they go by themselves and they grow by themselves. This is "the technological imperative"—what can be done must be done, and doing it leads to something else. In the process, the new replaces the old, even if the old was perfectly good. We see this all the time in the pharmaceutical industry: newer (and more expensive) drugs constantly are replacing other treatments or older drugs, often with minimal or no benefit compared with existing treatments.

Another characteristic of technological systems is their "totalization," their all-inclusiveness. In fact, says Ellul, there is really just one total technological system, and all subsystems (such as healthcare systems) are part of it. This super-system embraces all methods and techniques of all human endeavors (sports, medicine, art, religion, agriculture, education, manufacturing, and so forth) everywhere.

Finally, Ellul emphasizes that this technological system is autonomous; there is no person or committee or government able to hold it in check. In fact, no one is trying, because the benefits of the technological system are so pervasive. Few would want to halt the research for new "life-saving" drugs or surgical procedures. But even when questions are raised about the need for certain research and whether or not it really will lead to "life-saving"

---

9. Ellul, *Technological Society*, xxv.
10. Ellul, *Technological Society*, 79–147; Ellul, *Technological System*, 125–255.

procedures, there is no stopping the process: the momentum is too great. Yet this momentum itself points to a unique characteristic of the technological system: its lack of feedback. All natural systems have feedback loops that keep the system in homeostasis, that are self-correcting. However, the technological system has none: it only grows.

Unchecked growth: in healthcare, our biggest concern is growing costs; we are far less concerned about the growth of healthcare itself. In fact, in the USA we constantly talk of ways to *expand* coverage while keeping costs down. But unchecked growth is never benign. The essence of one kind of science fiction horror movie is an animal or insect that keeps growing. In biology, we call an unchecked growth cancer.

For those who want reform—either in financing this gargantuan system, or even in limiting its growth—we assume the change will come about through the political process. But again, Ellul brings us up short. The control we think we have to change things is an illusion. This illusion "falsely fulfills the need that individuals have for meaning, importance, effectiveness and security, leading them to surrender themselves all the more to the politicized state and the technicized system."[11] The Appendix to *Political Illusion* provides an excellent example of the folly of attempting to control economic planning by the political process; we would uncover the same folly if the example was reform of healthcare.

But how much of modern biomedicine itself is really useful, helpful, health-inducing? Consider people in the USA over sixty-five who are ill. Looking at Medicare expenditures spent on people who die in a given year, one-third of the money is spent during the last month of life. Most of these people were likely known to have terminal conditions, yet "the less money spent in this time period, the better the death experience is for the patient. . . . A lot of the money being spent is not only not helping, it is making that patient endure more bad experiences on a daily basis. The patient's quality of life is being sacrificed by increasing the cost of death."[12] This is an example of how the value of biomedicine in its present state is often an illusion, created by propaganda.[13] And because it's built on non-facts or selected facts, with the tools of propaganda, it becomes a matter for political action.[14] But the political action is hardly ever to challenge the essence of biomedicine; it is focused instead on the ephemeral matter of financing.

---

11. Fowler, "Synopsis and Analysis."
12. Bell, "Why 5% of Patients."
13. Ellul, *Political Illusion*, 96–135.
14. Ellul, *Political Illusion*, 103–12.

But neither size alone nor misplaced priorities are the biggest problem with biomedical healthcare. Rather, together they point to a deeper, more insidious problem—again one that Ellul emphasized throughout his writings: the problem of means becoming ends. The power of biomedicine is very real, and each of its specific offerings—a surgical procedure, a specific diet or exercise program, a medicine—may be quite helpful in relieving symptoms or removing a major problem such as a tumor or infection. These are all means. However, as we just saw with medical care at the end of life, biomedicine uses these means indiscriminately, without considering the ends. As in all technological systems, using the means has become the end in itself.

In biomedicine, this package of means is offered to everyone. But "offer" is too weak a word: Experts control the biomedical system, and most consumers do not have equivalent expertise to know whether to accept or reject what is "offered." It seems safest and easiest for most patients to simply accept the package of means, carrying the assumption that these means lead to the vague unarticulated end of "better health." In fact, as we saw with end-of-life care above, that end often remains unarticulated because is it does not exist. This means-become-end system is precisely what needs to be reformed, yet it is now beyond the ability of the political process to control.

So: Can USA healthcare be reformed? At most, the financing could be, following many European countries. But that is unlikely in our free enterprise system: Healthcare is a huge industry—some 16% of the GNP in the US. It is one of the main drivers of our national economy, and there would be little support for controlling costs and restraining this driver. Besides, the products of this system—technologies of symptom relief—are remarkably effective, *and* profitable. But financial reform would not challenge the technical nature of the means-become-ends system. Thinking that we could would be a political illusion.

Bibliography

Armah, Ayi Kewi. *The Healers*. Senegal: AFDJ, 2009.
Bell, Michael. "Why 5% of Patients Create 50% of Health Care Costs." https://www.forbes.com/sites/michaelbell/2013/01/10/why-5-of-patients-create-50-of-health-care-costs/#70f32d7328d7.
Ellul, Jacques. "Biblical Positions on Medicine." *The Ellul Forum* 59 (Spring 2017).
———. *The Political Illusion*. Translated by Konrad Kellen. New York: Knopf, 1967.
———. *The Technological Society*. Translated by John Wilkinson. New York: Knopf, 1964.
———. *The Technological System*. Translated by Loachim Neugroschel. New York: Continuum, 1980.

Fowler, James. "A Synopsis and Analysis of the Thought and Writings of Jacques Ellul." Christ in You Ministries. http://www.christinyou.com/pages/ellul.html.
Hadler, Nortin. *The Last Well Person: How to Stay Well Despite the Health-Care System*. Montreal: McGill-Queens University Press, 2004.
Illich, Ivan. "Brave New Biocracy: Health Care from Womb to Tomb. Life, Death and the Boundaries of the Person." *New Perspectives Quarterly* 11 (Winter 1994) 4–12.
———. *Medical Nemesis*. New York: Pantheon, 1982.
Illich, Ivan, and David Cayley. *The Rivers North of the Future: The Testament of Ivan Illich*. Toronto: House of Anansi, 2005.
Sontag, Susan. *Illness as Metaphor; and AIDS and Its Metaphors*. New York: Picador, 2001.
Sunga, Annette, et al. *Care of Cancer Survivors*. FP Essentials, Edition 352, AAFP Home Study, Leawood, Kansas, American Academy of Family Physicians, September, 2008.

18

The Environmental Movement in Ellulian Perspective

BY Jason Hudson

> *Jason Hudson is a PhD candidate at Cliff College (UK) and an adjunct professor at Cincinnati Christian University. His current work seeks to bring the works of Jacques Ellul and Wendell Berry into confrontation with current attitudes toward progress, particularly in the context of evangelical Christianity. He is particularly interested in the ways in which technological attitudes influence theologies of worship and approaches to ecology. He lives in Northern Kentucky where he seeks to live out his intellectual convictions through fidelity to his family, his community, and his acre of land.*

* * * *

IN A 1983 INTERVIEW with Berta Sichel, Jacques Ellul claims that he and his friend Bernard Charbonneau "began the ecological movement in France."[1] Christian Roy extends this claim by dubbing Charbonneau's manifesto, "Feeling for Nature as a Revolutionary Force," "the founding document of political ecology as an autonomous revolutionary position beyond Right and Left." Roy goes on to assert that it is "to this small regional stream of the French personalist tradition, that the origins of the ecological movement must be traced."[2]

But any such claim must be carefully parsed; the nineteenth century, for example, saw the rise of many well-known naturalist and movements hostile to industrialism. Nonetheless, Ellul's own unwavering commitment

---

1. Sichel, "New Hope for the Technological Society," 203.
2. Roy, "Ecological Personalism," 34.

to ecological activism and his important role in the ecological movement in France can be traced throughout his career. In the Sichel interview, Ellul reaffirms his commitment to ecological action even as he laments how the movement that he began had changed, stating that, "we're very critical of what the movement has become now. They lost the truth of their fight. But we are trying to recreate the movement."[3]

Those looking to Ellul for a positive ethic, a way forward with an easy conscience, will be disappointed at every turn. "I am critical of the French left," he says, "yet I am a leftist. I am critical of the Protestant church, yet I am a member of that church. I am critical in ecological circles. . . . All the groups that could listen back away."[4] The tension Ellul describes as he engages with groups and remains critical of them is directly related to his particular method. He strives, on the one hand, for an iconoclastic realism that exposes and destabilizes ideologies, even those that are at work within movements or institutions whose goals align with his own. Yet his dialectical approach to theology compels him to remain present with others, to join movements and so on, even as he critiques them. Ellul maintains a precarious tension: to be present in the group while not being fully assimilated into it, thus remaining a distinct other. Following Ellul's method, this essay seeks to navigate the tension of being present and deeply engaged in the environmental movement while exposing the ideologies at work in the movement, and in the space opened by this critique to propose an Ellulian approach to ecological activism.

Technician's Ecology

The environmentalism that Ellul laments and opposes is the technician's environmentalism. In *Living Faith*, he devotes a chapter—"We Never Wanted That..."—to reflecting upon the unintended consequences of well-meaning action. After decrying the proliferation of mediocre and clichéd books on the nature of technology that followed the publication of *La Technique*, Ellul writes, "The same thing goes for the French 'ecology' movement, which began as a prophetic message and became a sociological trend, a springboard for political success, and a way back into the industrial system. . . . The whole thing turned false when it was embraced by the public and the politicians, the media."[5]

---

3. Sichel, "New Hope for the Technological Society," 203.
4. Sichel, "New Hope for the Technological Society," 192.
5. Ellul, *Living Faith*, 216.

According to Ellul, the environmental movement ceases to be significant when it is subsumed by technological society. The current iteration of environmentalism demonstrates that it is, in fact, a technological movement by framing the ecological crisis in abstractions and as a primarily technical problem, by creating a political illusion meant to ease the conscience, and by using propaganda to create a mass mentality that shapes the psychology of the would be activist.

Technical Framing and Abstraction

The language of environmentalism exposes its technological underpinning in that it speaks in abstractions and frames the ecological crisis as primarily, if not exclusively, a technical problem. According to Ellul, "Technique, in its development, poses primarily technical problems which consequently can be resolved only by technique. The present level of technique brings on new advances, and these in turn add to existing technical difficulties and technical problems, which demand further advances still."[6]

Isabell Lamaud has demonstrated that this technical attitude is the underlying assumption of ecological modernization.[7] She notes the way in which the technical framing of ecological problems has led to the self-augmenting nature of technology: "destruction caused by the technician's system does not constitute a barrier to its further growth, but on the contrary can create new opportunities for expansion."[8] Though the language varies significantly across the political spectrum, the underlying assumptions remain that the ecological crisis is a technical problem in need of technical solutions. For the technological activist, time to consider de-growth or cultural and personal repentance as a response to industrial and technical growth has long passed. The problem has grown beyond human scale and can now only be addressed by technicians and their inevitable breakthroughs.

Moreover, this technician's environmentalism is articulated in abstractions. The word "environment" is itself an abstraction, since in fact there is no such place as an *environment*; there is no such *thing*. Therefore, it would seem repeated calls for environmental action cannot direct meaningful

---

6. Ellul, *Technological Society*, 92.

7. Lamaud carefully parses ecological modernization according to various scholars' conceptions of the term. She concludes, however, that the common factor among them is "the belief in the capacity of a technician form of management in the face of environmental problems." Here, then, ecological modernization is a stand-in for all neoliberal or growth-oriented approaches to ecology.

8. Lamaud, "Against Environment Protection?," 94.

action, but instead point only to vague sentiments and abstract gestures. Such abstraction may be useful as an umbrella term under which particulars can be referenced. However, this nuance is important as technology alienates people from concrete life. Abstractions ensure that individuals will not feel the gravity of the ethical dilemma in earnest as that would require a distinct and particular other to which one is responsible.

Continuing in the realm of abstractions, "the future" and forecasts have become the primary motivator for ecological action. Regardless of the quality of the forecasts, these are not current realities. For Ellul, the "mythic element inherent in the politicization of problems and values springs from the facility with which all things are relegated to tomorrow or the day after tomorrow."[9] He further explains the ethical implications of this abstraction: "And with admirable facility everybody avoids the personal question of his own conduct by politicizing it. The more the solution is in the future, the more everything is permissible today."[10]

Here again, the severity of the ethical situation is ameliorated with the abstraction of the particular ecological other. Wendell Berry argues a similar point: "If the future is threatened by the present, which it undoubtedly is, then the present is more threatened, and often is annihilated, by the future."[11] In short, as far as the technician's environmentalism creates abstractions, it leads away from the particular objects of concern and therefore cannot lead to meaningful activism but relegates all problems to the realm of technicians and experts.

Together, the abstract motivations for ecological action and the technological framing of the issue leave people feeling helpless. What can be done? However, those who believe the scientific consensus about climate change and feel the weight of ethical responsibility to leave a livable ecosystem for future generations must do something. This situation creates a tension and an anxiety that unmitigated makes life untenable. The technological system, however, has devices, human technology, to help ease the conscience of one who lives in this tension.

Everything Is Political

First, the relegation of environmental issues to the realm of politics and technology may allow the average person to feel an easing of ethical responsibility, since such a colossal and abstracted problem is well beyond

9. Ellul, *Political Illusion*, 189.
10. Ellul, *Political Illusion*, 190.
11. Berry, *Our Only World*, 174.

the scale of the single individual. When the problem is framed in global, abstract, and technical terms, the individual sees no way to take up the issue personally. The issue, it seems, is with multinational corporations and industrial nations.

The politicization of the ecological movement has resulted in a neoliberal approach to environmental protection, in effect, framing ecological crises as technical problems that can only be solved technologically. Lamaud's essay, "Against Environmental Protection? Ecological Modernization as 'Technician Ecology,'" demonstrates an Ellulian critique of neoliberal approaches to environmental protection. Neoliberal, pro-growth, approaches currently dominant environmental policy on the political Right and Left, since the working assumption is that pollution and ecological degradation are the by-products of costly corporate inefficiencies.

Thus, the thinking goes, businesses must have incentives to innovate in ways that will lead toward more efficient and cost-effective means of production, which will in turn create an *added* benefit of environmental health. For example, the carbon marketplace provides financial incentives for reducing pollution, making the "right to pollute" a commodity that can be exchanged or sold by ecologically savvy companies to those that are heavier polluters. This big-business model of responding to environmental problems unapologetically orients environmental protection within the domain of technology and efficiency. Ecological health is not posited as a value and cannot, therefore, act as a limiting factor to industrial or technological growth by its own virtue. Neoliberal platitudes about "competitive edge," "profit motives," "job creation," and the rest, mask what are merely conventional approaches to ecological health that perpetuate technology's autonomy and self-augmentation toward efficiency, to the detriment of human agency and ecological health.

Propaganda and Green Psychology

Another form of human technology helps individuals cope with this difficult situation by assimilating them into an increasingly technical and inhuman environment: namely, propaganda. Certainly, right wing propaganda that spreads doubt about the seriousness or reality of the ecological crisis is a serious problem. However, right wing propaganda and environmentalist propaganda are equally possible because most people remain utterly at the mercy of experts. This vulnerability creates easy targets propaganda's influence.

Because information is only accessible through technical intermediaries—through symbols and figures distilled from trusted authorities that shape facts and give them meaning—public opinion can be manipulated. According to Ellul, "public opinion always rests on problems that do not correspond to reality."[12] It does not arise out of people's real experiences, but from abstract notions presented in politically-charged symbols. It is important to see the nuance and difference between reality and truth. To say that public opinion does not correspond to reality is not the same as saying that it is false. Against popular conceptions, Ellul shows that propaganda is not primarily the propagation of falsehoods. He explains, "The truth that pays off is in the realm of fact. The necessary falsehoods, which also pay off, are in the realm of intentions and interpretations."[13] Environmental propaganda, likewise, is not intended to make a lie take hold; it is meant to use facts to create public opinion and motivate action. According to Ellul,

> The individual will never begin to act on the basis of facts, or engage in purely rational behavior. What makes him act is the emotional pressure, the vision of a future, the myth. The problem is to create an irrational response on the basis of rational and factual elements. That response must be fed with facts, those frenzies must be provoked by rigorously logical proofs.[14]

Environmental propaganda illustrates this use of facts to create a public opinion. This opinion, because it is shaped by symbols, creates consensus. Consensus creates in-groups and out-groups which gives members a sense of belonging and kinship as well as a common enemy, further ossifying the group's orthodoxy. Ellul explains,

> Propaganda attaches itself to man and forces him to play its game because of his overwhelming need to be right and just. In every situation propaganda hands him the proof that he, personally, is in the right, that the action demanded of him is just. . . . Propaganda appeases his tensions and resolves his conflicts. It offers facile, ready-made justifications, which are transmitted by society and easily believable.[15]

Environmental propaganda gathers people with this longing for justification and gives them a common language, a common enemy, a history, and a future. In short, propaganda creates unifying myths. According to Ellul

12. Ellul, *Propaganda*, 101.
13. Ellul, *Propaganda*, 53.
14. Ellul, *Propaganda*, 86.
15. Ellul, *Propaganda*, 158.

people in the technological milieu need these myths to connect to others. The myth assimilates people into the technological age and "keeps them from going mad."[16] Assimilation, when complete, creates groups of people who can be stimulated to action when the compression of facts and intentions into symbols—slogans and images—calls to mind the myth. Because the technician's environmentalism is subsumed by the technological society, it cannot spawn action that fundamentally challenges the assumptions of technology. Therefore, environmental propaganda can only birth action that is consistent with the technological system. These environmentalists can be called upon to vote for "green" candidates, purchase "green" products, and donate to "green" causes. However, so long as environmentalism remains a technical phenomenon it cannot go further than these gestures.

Agrarian Ecology[17]

Despite the myriad ways the ecological movement has been subsumed by the technological society, an Ellulian perspective does not allow for despair. Despite the common misconception of Ellul, his view emphasizes hope and action in the face of an admittedly daunting reality. Ellul was certainly critical of what the ecological movement had become by 1983[18]—what, then, did Ellul have in mind when said that he was "trying to recreate the movement"? Given his ethical commitment to freedom and his dialectical approach, Ellul's work is never overly prescriptive. Instead, he outlines parameters within which actions can be judged as more or less deterministic relative to their incorporation into the technological system. To develop a consistently Ellulian ecology, the implications of his thought must be extended beyond its original content, although such a method is directly supported by Ellul. His commitment to ecological activism in his own words, is

> fully in keeping with all my research on the technological society and the present influence of industry on the transformation of the human environment.... For me, it's always the same

---

16. Ellul, *Presence of the Kingdom*, 94.

17. Agrarian here is meant to be the English equivalent of *paysan* in French. Daniel Cérézuelle has noted the similarity between Bernard Charbonneau's paysanne ecology and Wendell Berry's agrarianism. Cf. Cérézuelle, "Wendell Berry et Bernard Charbonneau."

18. This date specifically references Ellul's previously mentioned interview with Berta Sichel. Other sources show that he was disillusioned with the movement prior to this interview.

problem. Intellectual interest means concrete commitment, practical and political involvement.[19]

According to Ellul, his criticism of the technological society must result in practice; however, application of Ellul's critique will never be monolithic, but will vary across time and place. But several factors will remain consistent: action must be local, rooted in people's concrete experiences, and the ethical burden of ecological stewardship must be shared by individuals; problems cannot be solved by mere abstraction and politicization.

Act Locally

Ellul began several groups interested in "the cause," though he is best-known for his work with the Committee for the Protection of the Aquitaine Coast, which opposed pollution and development of the Atlantic Coast near his home in Bordeaux. Ellul's own actions reflect his maxim, "think globally and act locally," and his concern for the environment in general took the form of protecting vulnerable regions nearby. While the technician's environmentalism describes problems as abstractions, agrarian ecology reunites activists through work in local landscapes, local food production, and shared common spaces. This emphasis on the local gives activists a meaningful path to action.

Rooting ecological action in particular localities, those in which activists live, work, and play, provides an advantage over technicians and experts who are often aloof and detached from nature. Because of this, these technicians tend to study ecology according to technological assumptions and industrial values which abstract particulars as "natural resources." The local activists, conversely, have access to local knowledge that connects particular landscapes and biota to a community over time. Local action that emerges from local knowledge can spawn passionate and effective protests. These activists care for the threatened environment because of their intimate knowledge of its importance for local culture as well as ecological and human health.

Rooted and Placed

Agrarian ecology places an emphasis on local action that rises from connections between people and their particular places and the interconnectedness of all places, particularly between cities and rural environments, which are

19. Ellul, *Presence of the Kingdom*, 20.

mutually interdependent. Bernard Charbonneau, Ellul's partner in ecological activism and intellectual study, rooted his ecology in agriculture and the lives of rural people. While Ellul doesn't take up this notion of a *paysanne* ecology directly, he commends it and suggests that Charbonneau's work is a direct correlate to his own. For example, in an interview, Patrick Troude-Chastenet asked Ellul about Charbonneau's influence on an *Espirit* document that cites the desire "to maintain man in contact with his neighbors and with a world where the only creative force comes from real life." Ellul responds, "Bernard and I were alone in advocating this particular political stance which later became the ecology movement. On the human level Bernard defended the position of the country people, the relation between man and nature and I addressed the economic consequences of emptying the countryside which I considered would be catastrophic."[20]

Ellul further explains that the breakdown of organic, localized, small groups and local institutions—those which were the backbone of traditional rural communities—is a condition for creating the kind of individualistic, mass society, which is primed for propaganda's influence. He writes,

> One can say, generally, that the nineteenth-century individualist society came about through the disintegration of such small groups as the family or the church. Once these groups lost their importance, the individual was left substantially isolated. He was plunged into a new environment, generally urban, and thereby "uprooted." He no longer had a traditional place in which to live; he was no longer geographically attached to a fixed place, or historically to his ancestry. An individual thus uprooted can only be part of a mass.[21]

For Ellul, then, urbanization and the disconnection between people and place is one of the conditions of propaganda and, more generally, a consequence of technology. Certainly, however, the key distinction between the isolated, individualistic person who is primed for propaganda and those who remain shielded, to a degree, by connection to a small group is the connection to a place and a community, whether rural or urban.

In fact, a great shift has occurred in the past century. Rural communities have become increasingly barren. The small institutions that once nourished rural communities are struggling or gone entirely. Their populations have diminished as farmers have "gotten big or gotten out."[22] Those who have

---

20. Troude-Chastenet, *Jacques Ellul on Religion*, 64.

21. Ellul, *Propaganda*, 92.

22. "Get big or get out" was a favorite maxim of Earl Butz, the agribusiness champion and USDA chief under Richard Nixon. In the decades after World War II

gotten big have replaced human labor with increasingly large machines and automation. Many who have gotten out have moved to urban centers in search of work. In the wake of this great migration, unemployment, depression, drug addiction, cultural stagnation, etc. has taken root. Conversely, these kinds of small, local institutions have strengthened in many cities, making it easier to connect through small groups. Even a concern for localized agriculture has flourished in many cities through community gardens, farmer's markets, and community supported agriculture.

An agrarian ecology that rises out of individual's connection to a particular place and the needs of that place is not limited to rural environments. Perhaps the key distinction that makes this ecology truly radical, even to the point of rupture in the technician's environmentalism, is that the particular needs of unique places challenge the universalism of the technological society. The eccentricities of place will influence the technologies employed in agriculture, development, manufacturing, and so on.

Personal and Cultural

Finally, agrarian ecology that is rooted in the real lives of people living in particular locations lays an ethical responsibility squarely on the shoulders of individuals. The removal of abstractions and ideologies leaves only the individual and the ecological other. Particularizing in this way creates an ethical situation in which action is motivated by affection for real creatures in real places. In this situation environmentalist ideology and slogans are irrelevant. First, they fail to address problems as experienced in particular localities; secondly, individuals are concerned with the health and wellbeing of the other more than being on the right side an issue. Agrarian ecology may not lead to an easy conscience, since problems cannot be simply turned over to big business and politicians to resolve, but through communal effort may lead to healthier environments. But the only way for agrarian ecology to ease the tension created by technological society is to challenge the progress narrative at its root and turn activists toward local realities to begin living *conservingly* in response to the needs of particular localities and communities.[23]

---

his maxim was realized: farmers left the business in droves and those who remained accrued debt to expand their fields and augment their machines.

23. This term seems to be an invention of Wendell Berry. He frequently uses it to describe a lifestyle that promotes long-term ecological flourishing.

Objections

My analysis here is subject to the same critique that much of Ellul's work invites. His emphasis on small, localized actions is likely to leave those concerned with the existential threat of global climate change wanting more. Agrarian ecology, however, does not negate the need for environmental protections on a policy level, new technology for energy creation, economic imagination for rethinking neoliberal, pro-growth models, and on and on. Instead, my argument here suggests that if the way we address environmental/ecological problems depends only upon technology and politics, then the underlying assumptions that have led to the ecological crises of today will continue and will perpetuate future crises.

We must acknowledge the truth of the addiction recovery mantra that our *best* thinking got us to this moment of crises, and that the current trajectory of technological expansion (today's best thinking) will be the genesis of tomorrow's crisis as well. Only a radical and critical ecology that questions the underlying patterns of development, and assumptions about what constitutes progress, can adequately address the problems confronting us today.

Some may object to the emphases on personal responsibility, suggesting that cultural change plays into the hands of neoliberal policy makers and industry. Again, the agrarian approach does not negate the need to hold industry leaders and policy makers to account. Indeed, rooting activism in connections to unique localities should spur action on the policy level, not supplant it. Others may object that this critique ignores the ways in which right-wing propaganda spreads false information and poisons the well of scientific consensus. Like Ellul, my intent has been to critique the environmental movement precisely because it is the position nearest my own. Of course, propaganda efforts from the right are pernicious and must be exposed.

Conclusion

Mass extinctions, droughts, top soil erosion, rise in oceanic temperatures, massive dead zones in the Gulf of Mexico, islands of floating plastic debris, nutrition-related epidemics in human health, and on and on, ad nauseum, clearly demonstrate that the world is facing an ecological crisis that rises to an existential threat. To allow the crisis to be subsumed by technology, framed as a technical problem to be addressed by technicians, is to surrender to the self-augmenting, autonomous technology that can only perpetuate future ecological crises regardless of its intent. Ellul sums up his objection to technological environmentalism saying,

> taking an interest in the protection of the environment and in ecology without questioning technical progress, the technician society, the passion for efficiency, is to undertake an operation which is not only useless, but fundamentally harmful. Because it will lead to nothing, but one will have the impression of having done something, it will allow one to falsely calm legitimate worries by throwing a new shroud of propaganda over the threatening reality.[24]

Ellul's ecological ethic demands nothing short of revolutionary action, action that truly strikes at the foundations of the technological society. Revolutionary action cannot be ossified into a dogma or a program, but each new situation requires imaginative and daring thinking and acting.

Bibliography

Berry, Wendell. *Our Only World: Ten Essays*. Berkeley: Counterpoint, 2015.
Daniel Cérézuelle. "Wendell Berry et Bernard Charbonneau." *Encyclopédie de l'Agora*, April 1, 2012. http://agora.qc.ca/documents/agriculture_biologique-wendell_berry_et_bernard_charbonneau_par_daniel_cerezuelle.
Ellul, Jacques. *Living Faith*. Translated by Peter Heinegg. San Francisco: Harper Row, 1983.
———. "Plaidoyer Contre la 'Défense de l'Environnement.'" Translated by Isabelle Lamaud. *France Catholique* 1309, 1310, 1311 (January 1972).
———. *The Political Illusion*. Translated by Konrad Kellen. New York: Vintage, 1972.
———. *The Presence of the Kingdom*. Translated by Olive Wyon. Philadelphia: Westminster, 1951.
———. *Propaganda: The Formation of Men's Attitudes*. Translated by Konrad Kellen and Jean Lerner. New York: Knopf, 1965.
———. *The Technological Society*. Translated by John Wilkinson. New York: Knopf, 1964.
Lamaud, Isabelle. "Against Environment Protection? Ecological Modernization as 'Technician Ecology.'" *Philosophy of Engineering and Technology* 13 (2013) 83–96.
Roy, Christian. "Ecological Personalism: The Bordeaux School of Bernard Charbonneau and Jacques Ellul." *Ethical Perspectives* 6 (1991) 33–44.
Sichel, Berta. "New Hope for the Technological Society: An Interview with Jacques Ellul." *Et Cetera* 40 (Summer 1983) 192–206. http://www.jstor.org/stable/42576602.
Troude-Chastenet, Patrick. *Jacques Ellul on Religion, Technology, and Politics*. Translated by Joan Mendes France. Atlanta: Scholar's Press, 1998.

24. Ellul, "Plaidoyer Contre," 11.

19

Charlottesville vs. the Real Revolution

BY Michael Morelli

> *Michael Morelli is a PhD candidate studying theological ethics and technology at the University of Aberdeen, Scotland. He is originally from Vancouver, Canada, where he received a BA in English and Communications (Simon Fraser University) and an MA in Christian Studies (Trinity Western University). He has also worked in local churches and is a freelance writer.*

* * * *

CITIES AND REVOLUTIONS OFTEN come in pairs but not all revolutions in the city are welcome or constructive. A paradigmatic contemporary example is the "Unite the Right" protests and "Anti-Fascist" counter-protests which occurred in Charlottesville, Virginia on August 11 and 12, 2017. During the evening of August 11, approximately two hundred fifty people gathered at the University of Virginia campus to protest the removal of civil war monuments representing confederate leaders. Protestors wore khaki trousers and polo shirts, held tiki-style torches, and marched across campus chanting *"blood and soil," "you will not replace us," "Jews will not replace us,"* and *"white lives matter."* The protestors received extensive media coverage and it was a very public performance of various white supremacist and white nationalist groups which had organized online. Violence erupted when protesters were challenged by a small number of counter-protesters who were mostly University of Virginia students. Confrontation ceased when police intervened but much more violence would bleed out into the next day.

On the morning of August 12, white nationalists and white supremacists gathered at Charlottesville's Emancipation Park and Market Street. Many protesters wore armor and carried banners, clubs, shields, and firearms. Counter-protesters were also present and many wore armor and masks and carried signs, shields, bludgeons, and pepper spray. Amidst the protesters and counter-protestors was a small contingent of non-violent demonstrators including clergy, people of faith, civil rights leaders, and impartial onlookers. Local police were also stationed at the protests but various reports state they did not make a significant effort to prevent the physical altercations that took place—even when one protestor, a white supremacist, discharged a firearm in the direction of counter-protestors. Fortunately, nobody was hit by that bullet.

With each of these groups at this protest representing different segments of society—politically "right," politically "left," religious groups, human rights advocates, the state power of the police, and fringe and mainstream media organizations—everything exploded. The significant civil unrest and alleged "revolution" and "protest" over the course of these two days reached its highest pitch when a white supremacist intentionally drove his Dodge Challenger into a crowd of counter-protesters. One person was killed and nineteen were injured among the many other people injured by the various physical altercations that took place.[1]

There is much to reflect on in the aftermath of Charlottesville, especially in view of the fact that events like these are increasingly characterizing "political expression" in North America. Of specific interest to this chapter is how there has been limited discussion in mainstream media about the ways in which technology connected these various white nationalist and supremacist groups online, enabled them to assemble in a city offline, and broadcast to the world their actions.[2] Would this destructive and senseless revolt in an unsuspecting city have occurred at the level of chaos and violence it did were it not for the power afforded to people by technology? And where were the Christians when these events transpired? There was a small number of clergy present at these events and we will describe and reflect on their actions below.

For Christians who watched these events unfold with horror, it was shocking to realize that among the crowd chanting "*blood and soil*" and "*Jews will not replace us*" were people who self-identify as "Christian." Remember that after terrorist attacks committed by self-identified Muslims, many voices demanded a strong disavowal and rejection of terrorism and violence by Muslim leaders with assurance that this was not a part of basic

---

1. See Heim, "Recounting a Day of Rage," for a detailed time line of events.

2. Fang and Woodhouse, "How White Nationalism Became Normal Online," is one of the few thoughtful and well-researched journalistic articles that have reported on this subject.

Islam. Should the same not be demanded now of Christians?[3] The fact is there *were* self-identified Christians present in the Charlottesville group promoting white supremacy and I would contend that not only should Christians repudiate what the white supremacist demonstrators believed, said, and did but that Christians theologians have a responsibility to explore why this aberration from basic Christian faith developed and provide a counterpoint.

Radicalizing Technology

In *Autopsy of Revolution* Jacques Ellul describes how commonplace technologies have reified revolution in the modern world:

> We are made to feel (on the emotional rather than the intellectual plane) that we live in a revolutionary world. Little by little we have come to believe that revolution is besieging every corner of our universe, that it is the permanent reality of life, and that it is everywhere and accounts for everything. Does this mean that we are riding a great wave of revolution? It means just the reverse: that we have reduced revolution to the level of daily news, along with winter sports, the newest automobiles, the latest presidential press conference, the miniskirt, and the Goncourt literary award, and that it ranks equally with all of them.[4]

This is why Ellul argues that much of what masquerades as "revolution" today actually is explosive but ephemeral "revolt."[5] What took place in Charlottesville was a *revolt*. It was not a *revolution*.[6] It did not change anything in any enduring way. It only revealed the depth and directionlessness of North American malaise.

Although Ellul's observations were published in the early 1970s and therefore shaped by a different sociopolitical context, it is striking how pertinent they are forty years later. In fact, it is arguable that these analyses have become more accurate given the degree to which today's technologies have flattened, fragmented, and complicated revolution beyond the technologies

---

3. I am basing this supposition on reliable journalistic reporting such as Worthen, "Match Made in Heaven"; Schmidt and Wang, "Jerry Falwell Jr."; and Green, "How Will the Church Reckon With Charlottesville?" There were many President Trump supporters in the crowd of protestors and it is an established fact that President Trump won the approval and votes of many evangelical Christians during his political campaigning.

4. Ellul, *Autopsy of Revolution*, 175.

5. Ellul, *Autopsy of Revolution*, 8–9.

6. Ellul, *Autopsy of Revolution*, 38–39.

of Ellul's time. When Ellul wrote *Autopsy*, the public understanding of what constituted revolution was almost entirely shaped by a limited number of mainstream media. In our time, almost anyone can create a "news platform," gain an audience, argue a position, and incite actual revolt. This is evidenced by the rapid development and proliferation by individuals or groups, of alternative media, which can radicalize people and activate violent revolts such the ones which took place in Charlottesville.[7]

There is significant research and discussion about how communication technology is used to radicalize Muslims. But it is equally important to initiate similar discussions about how such technology is being used to radicalize Christians, as well as white middle- and working-class North Americans.[8] Dylann Roof, who killed nine people in the Emmanuel African Methodist Episcopal Church, was a member of a Lutheran church, identified as Christian, and was radicalized online.[9] But the question is: *how* was a person like Dylann Roof radicalized?

Here Ellul's descriptions of propaganda and its relationship to technology are helpful:

> Propaganda tries to surround man by all possible routes, in the realm of feelings as well as ideas, by playing on his will or on his needs, through his conscious and his unconscious, assailing him in both his private and his public life. It furnishes him with a complete system for explaining the world, and provides immediate incentives to action. We are here in the presence of an organized myth that tries to take hold of the entire person. Through the myth it creates, propaganda imposes a complete range of intuitive knowledge, susceptible of only one interpretation, unique and one-sided, and precluding any divergence.[10]

This suggests how, for Dylann Roof and for many of the protestors in the crowds at Charlottesville, propaganda coupled with technology facilitated total submersion in mythologies which galvanized into violent action in the heart of a city. In these cases, mythologies were brutally charged notions of *whiteness, patriotism, nationalism, freedom, religion, revolution,* and the like, and in filling these signifiers with hateful meaning, propaganda used commonplace technologies to form mythologies *of* revolution which provoked people *into* violent revolt.

---

7. Glaser, "Internet of Hate."
8. Manjoo, "Hunt for Ways."
9. Kaleem, "South Carolina Lutheran Pastor"; and Quinlan, "Dylann Roof."
10. Ellul, *Propaganda*, 11.

Ellul stresses that propaganda in its modern form must be understood within the context of technological society. Propaganda is more than deceptive posters, speeches, and broadcasts inciting people to action. Propaganda still takes these obvious forms, of course, but Ellul argues that it is more fully understood as the means by which individuals and groups are integrated into and activated in a society increasingly technological and complex. "Propaganda is called upon to solve problems created by technology, to play on maladjustments, and to integrate the individual into a technological world." "Propaganda must be situated at the center of the growing powers of the state and governmental and administrative techniques."[11]

What took place in Charlottesville was shaped by societal maladjustment activated by communications technology and propaganda. But this does not absolve people from responsibility as though they were simple cogs in the machine. Propaganda is not just unidirectional; people actually *desire* propaganda. It "corresponds to deep and constant needs, more developed today, perhaps, than ever before: the need to believe and obey, to create and hear fables, to communicate in the language of myths. It also responds to intellectual sloth and desire for security."[12] The myth of *blood and soil* is one such myth which capitalizes on and violently activates intellectual sloth and desires for security.

At the same time, we cannot consider propaganda on the fringes of society to be separate from propaganda at the center of society. Propaganda is reliant upon a technological network with different nodes that connect to, reinforce, and radicalize each other.[13] Ellul argues that propaganda cannot create mythologies out of nothing in a society, but can only radicalize mythologies which already are present.[14] Candidate and President Trump's *Make American Great Again* slogan wouldn't work if it didn't correspond to the grievances as well as aspirations already harbored by Religious Right white Evangelical Christians. The chants of *blood and soil* connected to an already-existent anti-Semitism, racism, and fear of the "other" shared by many.[15] Social groups, with their attendant mythologies, are mobilized by propagandistic messages and symbols that reinforce

---

11. Ellul, *Propaganda*, xvii–xviii.
12. Ellul, *Propaganda*, 148.
13. A full account is given in Ellul, *Technological System*.
14. Ellul, *Propaganda*, 33.
15. It is telling in this regard that President Trump initially did not directly condemn the neo-Nazis or Klu Klux Klan protestors in Charlottesville. Whether this was due to fear of offending or alienating supporters represented in these groups, whether he felt some of the protestors reflected views of his own, or it was a combination of both, remains up for speculation. For example, see Hasan, "Donald Trump."

identity and meaning. Scapegoats and enemies then become the targets of terrible verbal and physical violence.

The Reservoir of Action

So what are we to think of clergy and believers at these events? First, let's look at the context of the actions of the Christian clergy and people of faith in Charlottesville. Cornell West, a prominent clergyman, academic, and activist at the event, describes what happened as follows:

> I was going to hear a powerful sermon by my dear sister Reverend Dr. Traci Blackmon. . . . [Protestors] held us hostage in the church [on Friday night]. We could not leave after the service, because the torch march threatened the people who were there. . . . The next day, [twenty] of us who were standing, many of them clergy, we would have been crushed like cockroaches if it were not for the anarchists and the anti-fascists who approached, over three hundred, three hundred fifty anti-fascists. We just had twenty. And we're singing "This Little Light of Mine."[16]

Other accounts confirm this description, including what seems like a very limited and modest nonviolent demonstration by singing at the center of a violent scene of marching, shouting, jostling, pushing, shoving and worse protest and counter-protest.

As limited as the actions of West and his clergy friends may seem, they in many respects capture Ellul's theological concept of the "violence of love."

> It is not by sequestering ourselves in our churches to say little prayers that we fight, but by changing human lives. And it is truly a *fight*—not only against our own passions and interests and desires, but against a power that changed only by the means which are the opposite of our own. Jesus overcame the powers—of the state, the authorities, the rules, the law, etc.— not by being more powerful than they but by surrendering himself even unto death.[17]

Ellul describes the counterpart to the "violence of love" in *Meaning of the City* where the city is the spiritual manifestation of humanity's technology, power, revolt, and violence. The founder of the city in the biblical narrative

---

16. DemocracyNow.org, "Cornel West & Reverend Traci Blackmon."
17. Ellul, *Violence*, 160 and 167.

is none other than Cain who murdered his brother Abel, and rather than accept his life as a wanderer became the first builder of a city.[18]

For Ellul, all of humanity has inherited the destructive spirit of Cain. Like Cain, we are self-serving builders and we incline to violence. Our hope is not to use the weapons of the enemy to defeat the enemy; to try to out-technique, technique, or out-propaganda propaganda. Hope is found in a radical break, in introducing the "wholly other" into the world, in something as unlikely and seemingly ineffective as singing "This Little Light of Mine" in the midst of a violent protest and revolt. Hope is "the word we have constantly found spoken to the city, containing both judgment and grace, [the] word of the cross."[19] The Gospel proclaims that Jesus Christ defeated the violent, oppressive powers of the world not with counter-violence, technology, or propaganda, but by his sacrificial death and subsequent resurrection. His act of non-power and love triumphed over the powers of the world. The grave could not hold him.

Hope, for Ellul, finds a tangible expression in prayer. Prayer shuts out the noise and power of technology and propaganda, and the incitement to violence.

> When modern man undertakes an operation he relies on those means which have long been shown to be effective. He does not rely on [God]. The results he is after are perfectly obtainable without prayer. This is so with increasing frequency. Hence prayer [is] fading out. We might even say that it is being made ridiculous by technology. . . . Prayer is ridiculed because its effectiveness is entirely unpredictable, and statistical techniques are able to show that the percentage of "answers" to prayer corresponds exactly to the percentage of success which would have been the case had events been allowed to take their own course, and without prayer.[20]

Indeed, in a technologically locked-up world, prayer looks to the world like one of the most pathetic acts possible. Why pray when we can use this or that technology to organize, change, influence, and revolt? Ellul argues that the most consequential act in a technological and propagandized society is precisely the act which society deems the most inconsequential; that the Christian who cannot pray in and for the city will be overrun and dominated by the city with all its technology and propaganda. Prayer puts one in communion (and in a "wrestling") with a God who commands "Thou shalt

---

18. Ellul, *Meaning of the City*, 6–9.
19. Ellul, *Meaning of the City*, 82.
20. Ellul, *Prayer and Modern Man*, 79.

not kill"—or even insult or hate—an enemy, a Lord who teaches that it is the child, the widow, the poor, the powerless, the marginalized who represent and inherit the kingdom of heaven, a God who insists that we should not repay evil for evil but overcome evil with good.

In *The Politics of God and The Politics of Man* Ellul writes with forceful irony, "Yes, prayer is useless, and so too are miracles and theology and the diaconate and works and politics." But, he clarifies, "It is just because these acts were useless and did not carry with them their own goal and efficacy that they are on the one hand testimonies to grace and on the other an expression of freedom. To be controlled by utility and the pursuit of efficacy is to be subject to the strictest determination of the actual world."[21] This is a theological inversion of a sociological reality. In a world that only values or permits the useful, the quantifiably effective, the most radical revolution springs out of the unquantifiable value of something useless and ineffective according to the standards of technique.

It is not that Christians will never be called to involve themselves in politics or revolution. Rather, it is that Christians are called to a different kind of presence in the world, a presence that flows from the reservoir of the practices of discipleship and the life of faith, hope, and love. Prayer will appear to be impractical and ineffectual to those who prioritize *doing* over *being* but prayer signifies the church's true hope for initiating a meaningful revolution in the city.

Bibliography

DemocracyNow.org. "Cornel West & Reverend Traci Blackmon: Clergy in Charlottesville Were Trapped by Torch-Wielding Nazis." *Democracy Now*, August 14, 2017. https://www.democracynow.org/2017/8/14/cornel_west_rev_toni_blackmon_clergy.

Ellul, Jacques. *Autopsy of Revolution*. Translated by Patricia Wolf. New York: Knopf, 1971.

———. *The Meaning of the City*. Translated by Dennis Pardee. Eugene: Wipf and Stock, 2011.

———. *The Politics of God and the Politics of Man*. Translated by Geoffrey W. Bromiley. Grand Rapids: Eerdmans, 1972.

———. *Prayer and Modern Man*. Translated by C. Edward Hopkin. New York: Seabury, 1970.

———. *Propaganda: The Formation of Men's Attitudes*. Translated by Konrad Kellen and Jean Lerner. New York: Vintage, 1975.

———. *The Technological System*. Translated by Joachim Neugroschel. New York: Continuum, 1980.

---

21. Ellul, *Politics of God*, 196–97.

———. *Violence: Reflections from a Christian Perspective*. Translated by Cecelia Gaul Kings. New York: Seabury, 1969.

Fang, Lee, and Leighton Akio Woodhouse. "How White Nationalism Became Normal Online." *The Intercept*, August 25, 2017. https://theintercept.com/2017/08/25/video-how-white-nationalism-became-normal-online/.

Glaser, April. "The Internet of Hate." *Slate*, August 30, 2017. https://slate.com/technology/2017/08/the-alt-right-wants-to-build-its-own-internet.html.

Green, Emma. "How Will the Church Reckon With Charlottesville?" *The Atlantic*, August 13, 2017. https://www.theatlantic.com/politics/archive/2017/08/will-the-church-reckon-with-charlottesville/536718/.

Hasan, Mehdi. "Donald Trump Has Been a Racist All His Life—And He Isn't Going to Change After Charlottesville." *The Intercept*, August 15, 2017. https://theintercept.com/2017/08/15/donald-trump-has-been-a-racist-all-his-life-and-he-isnt-going-to-change-after-charlottesville/.

Heim, Joe. "Recounting a Day of Rage, Hate, Violence, and Death." *Washington Post*, August 14, 2017. https://www.washingtonpost.com/graphics/2017/local/charlottesville-timeline/?utm_term=.62486c380ca1.

Kaleem, Jaweed. "South Carolina Lutheran Pastor: Dylann Roof Was Church Member, Family Prays for Victims." *Huffington Post*, June 19, 2015. http://www.huffingtonpost.com/2015/06/19/dylann-roof-religion-church-lutheran_n_7623990.html.

Manjoo, Farhadd. "A Hunt for Ways to Combat Online Radicalization." *New York Times*, August 23, 2017. https://www.nytimes.com/2017/08/23/technology/a-hunt-for-ways-to-disrupt-the-work-of-online-radicalization.html.

Schmidt, Samantha, and Amy Wang. "Jerry Falwell Jr. Keeps Defending Trump as Liberty University Grads Return Diplomas." *The Washington Post*, August 21, 2017. https://www.washingtonpost.com/news/morning-mix/wp/2017/08/21/liberty-university-graduates-return-diplomas-because-of-support-for-trump-by-jerry-falwell-jr/?utm_term=.d386f688668e.

Quinlan, Casey. "Dylann Roof Was Not 'Self-Radicalized.' He Was Part of a Racist Community." *Think Progress*, December 21, 2016. https://thinkprogress.org/dylann-roof-white-collar-white-supremacy-c548a1c20bef/.

Worthen, Molly. "A Match Made in Heaven: Why Conservative Evangelicals Have Lined Up Behind Trump." *The Atlantic*, May 2017. https://www.theatlantic.com/magazine/archive/2017/05/a-match-made-in-heaven/521409/.

Part Three: Appropriations

THE FINAL FOUR CHAPTERS suggest ways of appropriating Ellul's political thought in our own time and place. Patrick Troude-Chastenet looks at today's political situation in Europe and the United States and finds Ellul's warnings about the rule of the technicians beneath the society of spectacle completely apt (chapter 20). He offers little hope but certainly some realism and clarity of vision are essential if any real change is ever to happen. Doug Hill's essay (chapter 21) puts the focus even more on the propaganda which creates a "sham universe." Hill ends by a wise counsel to take a critical look at our own communication and not just condemn that of the opposition.

Adrian Almazan's essay on the "necessary revolution" (chapter 22) sketches out the enormity of the challenge of political and social change but provides some interesting hope. Weber (and Ellul's) "iron cage" of rationalism and technique, as dominant and oppressive as it is, does not lead inexorably to a utopian "brave new world" but in critical ways cannot completely conquer the mystery, dysfunction, randomness, and even "cussedness" of the world. Cracks appear in the walls, so to speak. And in those cracks we may still find the revolution we need.

And finally Andy Alexis-Baker's essay (chapter 23) describes a political-social-cultural landscape about as bleak as one could imagine. And yet, in an eloquent conclusion Andy recalls the alternative meta-narrative of the Hebrew prophets, which is no pie-in-the-sky comforting fantasy or illusion, but the powerful truth of a way that, while it is "wholly other," can revolutionize our here-and-now.

20

Democracy Confiscated?
The Continued Relevance of Jacques Ellul[1]

BY Patrick Chastenet

> *Patrick Troude-Chastenet is Professor of Political Science at the University of Bordeaux. He was Jacques Ellul's student assistant in the 1970s. He has published ten books, including* Lire Ellul *(1992),* Jacques Ellul on Politics, Technology, and Christianity *(2005),* Jacques Ellul penseur sans frontières *(2005), and* Comment peut-on (encore) être ellulien au XXIème siècle? *(2014). He is the founding president of the Association Internationale Jacques Ellul, director of Cahiers Jacques Ellul, and a founding board member of the IJES. He has organized several international multidisciplinary colloquia on Ellul's thought and legacy.*

* * * *

ELLUL WROTE THE CORE of his work on politics and policy from the Fifties to the Eighties. Is it out-of-date, or still pertinent? I will argue that, if anything, it is more relevant than ever, provided that we don't try to turn Jacques Ellul into some kind of Nostradamus who predicted everything before his death: from the subprime mortgage crisis to the latest electronic gadgets, and from the 9/11 terrorist attacks to the commercial failure of Google glasses!

---

1. Thanks to David Smith and Lisa Richmond for help on my English translation of this paper.

The Twin Rule of the Necessary and the Ephemeral?

Ellul's analysis of the "political illusion" has never been more relevant. "Everything is politics." But true politics, the science of choosing goals for society and the strategies to attain them, is an illusion in a technological society, a society immersed in a universe of sounds and pictures, dominated by what Ellul calls "the Necessary and the Ephemeral." The Necessary excludes choice—and choice is the essence of political action. The French Prime Minister Pierre Mendès France said "to govern is to choose." Ephemeral, on the other hand, describes how our political leaders act, focused on the short-term and changing their stance as often as the winds change. In the fifty years since Ellul, the world has become ever more *technicized* and "spectacularized." (When Shakespeare said, "All the world is a stage," he didn't imagine it would go so far!) The world has become a hyper-connected network, and images lie at its heart.

This makes Ellul's analysis of politics in a world of images even more relevant today than in 1965 when he published *The Political Illusion*. It is exemplified by propaganda shots from around the world—such as the famous footage of President Barack Obama, Secretary of State Hilary Clinton, Vice President Joe Biden, and several other key leaders, intently gathered around a screen in the White House Situation Room watching the capture and execution of Osama bin Laden *live!*

For Ellul, true political choice is based on total freedom: the ends come before the means. To paraphrase Hegel, a truly responsible political leader must be able, through his actions, to make necessity a contingency, and contingency a necessity. This is increasingly rare today. Take the example of Bin Laden: first, in May 2011, what was called "Operation Geronimo," the decision to capture him dead or alive, was a Necessity, not a choice, for whoever was the US President. Second, world-wide hype promoted this political "success" (although it took ten years for the world's greatest superpower to kill its Public Enemy #1). This is a perfect example of the "political show" foreseen by Ellul fifty years ago. The goal was to persuade international public opinion, starting with America, that the USA really was the leader. With the US government, Hollywood movies and super heroes are never far away. Nor is propaganda!

Across the Atlantic, Europe is bogged down by the question of how to deal with the flood of immigrants and refugees, and the European Union is powerless to stem a rising populism. Photos of French president François Hollande and German chancellor Angela Merkel in tender (but safely nonsexual) embrace, while gazing lovingly into each other's eyes, were aimed not so much at the commemoration of the centennial anniversary of the battle of Verdun as at broadcasting the subliminal message of a reconciled, pacified, and well-governed Europe.

On the other hand, Ellul thought that without the will to work for the future, there could not be any worthy policies. "Continuity is a human right" said a nineteenth century French precursor of John Rawls, Charles Dupont-White. Yet, Ellul shows that freedom of choice and control over the long term has vanished completely from our technological societies. This diagnosis—first formulated in 1965—is even more accurate today. The challenges faced by political leaders—nuclear proliferation, Islamic terrorism, the wars in Iraq, Syria, and the Sudan, the lightning-quick spread of viruses in a globalized world, food shortages and malnutrition, climate change, the pollution of the seas and waterways, the dangers of the nuclear industry, and all the problems related to energy supply and use and more generally to the very future of our planet—all these challenges require very long term analyses and highly ambitious solutions. Our political leaders compensate for their actual impotence by a kind of frenetic activity.

A look at the way the media are handled at the Group of Twenty (G20) summit meetings is enough to be convinced. In the area of financial (de) regulation, the continual summits, each one referred to as the "last chance," constitute the archetypal example. Ritualized and systematically presented as decisive moments in the strongest historical sense, these meetings try to save the appearances but cannot mask a reality that should depress not only the leaders but also us, the led. Today's European leaders depend on the will of the rating agencies, what Ellul called "l'aristocratie technicienne," in this case, a specific category of financial technicians.

Worse still, in the case of the Greeks: the troika of the European commission, the European central bank, and the International Monetary Fund forbade the Greek Prime Minister from submitting the question to a popular vote, and then simply ignored the "no" result of the 2015 referendum. In the name of the single currency and budgetary rigor, and by successive strokes of austerity plans, the "good governance" experts continued to strangle the heirs of Pericles and Athenian democracy. In the face of the international financial bodies, what is left of popular sovereignty and the democratic model? Is it not a perfect example of the Ellulian theory of bureaucracy and technocracy hollowing the value out of politics? Necessity rules!

As Ellul wrote in the aftermath of World War II in the protestant newspaper *REFORME*: "The Marxist analysis that reveals democracy to be a camouflaged dictatorship of the ruling classes is rigorously exact." As Marx said, "democracy is the means for the people to choose who will strangle them!"[2] The cases of Hungary, Poland, Finland, Austria, the United Kingdom, and France could lead us to believe that from now on the only choice of the peoples of Europe will be to arbitrate between the technocrats and the populists. Italy decides between the likes of Marco Dragi or Matteo Renzi against the likes of Sylvio Berlusconi or Beppe Grillo. As for the USA, it decides between the likes Hillary Clinton and Donald Trump. In such circumstances, was Ellul right to assert in the Sixties that democracy had already long since vanished?

Democracy: End, Decline, or Transformation?

We can blame Ellul for his idealistic approach toward democracy, stemming from theoretical confusion between democracy as an empirical regime, and thus necessarily imperfect, and democracy as a normative ideal to strive for, and thus continually reinvented. We would thus fault Ellul for having distinguished too little between the types of democratic regimes, each with its imperfections and inconsistencies, and for not having said loudly enough that they are preferable to the authoritarian or totalitarian alternatives. But we know that Ellul, in all fields, whether political or spiritual, saved the harshest criticisms for his own camp.

Given all this: is it right to say with Colin Crouch that democracy is disappearing, or is at least in steep decline?[3] A fundamental premise of the democratic model is that the sovereign people freely choose and oversee their representatives, who actually govern, giving orders to an administration completely obedient to the legitimately elected political power. But every day brings evidence in nearly all fields that the democratic model has been turned upside-down. Ellul of course told us that popular sovereignty is but a founding myth, and that there was never really any such thing. Ellul further clearly described increasing globalization when he underscored the "universalization of technique" that would make the world increasingly united, but increasingly controlled by an aristocracy of technicians, a caste coming from the finest schools. In France, these include the Grandes Ecoles like Polytechnique, ENA, Centrale, Normale Sup, l'Ecole des Mines de Paris, but also the Head Engineers of the Power Companies (EDF), the Atomic Energy Commission, the nuclear power plants, and the arms industry.

Ellul didn't use the word "globalization," but the idea is already there in his trilogy on technology. In our world united by technical rationality, who truly rules? Neither the people nor their representatives, but a third figure:

2. Ellul, "Quand les mots perdent leur sens," ("When words lose their meaning").
3. Crouch, *Post-Democracy*.

the technocrat, or rather an organization of unelected experts. In practically every facet of the lives of our communities, the real power is held by high level administrators and executed by their subordinate technical experts. It is they who deliberate and plan and make the key decisions. The parliaments, ministers, and presidents then make them their own and then sell them to the people through the news, which is simply good old-fashioned propaganda with a more socially acceptable dressing. This field has its own expert technocrats, the Spin Doctors.

The technocrats' strengthening grip on political choice destroys the very principle of popular sovereignty. A passage in *L'illusion politique* shocked me profoundly in 1977 when I read this book for the first time. "*Imbéciles qui croyez défendre ce qui a disparu depuis longtemps!*" (Fools who believe that they are defending what vanished a long time ago!). Looking back, we see that democracy as a political regime is spreading throughout the world, at the same time that it is evaporating as a political ideal. Who can seriously assert that political power ultimately belongs to the *demos*, and not to the aristocracy of technicians considered by Ellul to be the new ruling class? Just like the ancient aristocracies, as Ellul wrote in *le Bluff Technologique*, this class has exclusive knowledge, certain caste behaviors, and a jargon impenetrable by the common man. The class has access to and control of the centers of power. The members of the class subscribe to an ideology causing them to disdain those not of their clan, such as ecologists and ordinary folks.

The aristocracy is composed of high level civil servants and super-engineers, as well as the best financial technicians. Algorithm and digital experts play key roles, presumably soon to be joined by climate *geo-engineering* wizards. In the end, democracy may not have disappeared but it has massively mutated to what is not strictly speaking a technocracy: "political parties are not made up of technicians." Rather, we will call it a *technodemocracy*, a hybrid regime combining technocratic elements (legitimation by the *technè*—"the technically competent") as well as democratic elements (legitimation by the *demos*—by popular election). More and more, consulting "experts" reinforce the "technicist" ideology to the detriment of the voice of the citizen and, in the end, to the sovereignty of the people.

Experts Against Democracy?

The Expert is at the heart of the technological society. Defined simply by Ellul as "the best specialist on a given scientific subject," the *expert* fills the role of technology assessment, that is, he who explains to the public but who also shields from view that which must remain hidden. Traditionally, experts were neutral, independent of vested interest, bound to speak the full truth. It is no longer so. Not only does he plead for his organization or sponsor (agrochemical, oil, shale gas, nuclear industry, drugs, chemistry, digital economy, wind energy, telecommunications and so on) but the expert is also the best propagandist for the technical ideology.

For example, everyone in France remembers the Call from Heidelberg, signed by hundreds of doctors, intellectuals, writers and leading scientists including seventy-two Nobel Prize laureates, all "experts" in their fields, from Theatre of the Absurd to quantum physics. Issued on the eve of the Earth Summit in Rio, June 1, 1992, their text denounced the emerging "pseudo-science" of Ecology. It warned government leaders against these would-be protectors of the environment, saying that they were driven by an *"irrational ideology opposed to scientific and industrial progress."* It took twenty years to prove that there was nothing spontaneous about these experts' action. A Parisian firm lobbying for the asbestos and tobacco industries had deftly orchestrated the entire propaganda campaign.

This is not the place to detail my research on two French Public Health scandals that support Ellul's analysis.[4] One was the case of HIV-contaminated blood knowingly being distributed to hemophiliacs for several years; the other was the drug Mediator being prescribed to millions in spite of broad evidence of very serious side effects. (Mediator was called Redux in the USA) In both cases, political leaders were hostage to the Experts. We were wrong to laugh at the minister Georgina Dufoix when she said that she was "responsible but not guilty." Au contraire! Twenty-five years later her words show how the relations between leaders and experts have evolved. A minister determines health policy, but he is inadequately informed by the experts hired to guide him. Misinformed? Then "Not guilty!" But responsible because of the office he holds.

Is the State Still Enemy Number 1?

Ellul's critique covered both phenomena, the "technicization" of our societies as well as the central government's increasing control of society. Technology-worship and the State-as-Moloch were seen by Ellul as two sides of the same coin. A personalist in his youth, Ellul wanted the central government to have less control. He wanted to build networks of horizontal power, with grass-roots ecological regional projects growing into cooperation at the

---

4. Troude-Chastenet, "From the 'Contaminated Blood Affair,'" 145–58; Troude-Chastenet, "Santé publique et démocratie," 185–96.

federal level. He was in fact close to the anarchists, as readers of REFORME would guess in reading his post-war articles. He confirmed this forty years later in his book *Anarchy and Christianity*.

Is the State Still Growing Today?

Isn't Ellul's anti-State attitude very French? He is steeped in the State's role in the birth of the French nation, embodied in its constitution as it evolves and persists. Unlike Holland and Great Britain, in France it was not merchant companies that explored, conquered, and colonized. French history has always been top-down, from Paris. French political culture is traditionally centralized and Jacobin, a religion of public service in the citizen's best interest. Individual citizens, real people out in the provinces, are forever suspected of self-interested and personal goals that are contrary to the assumed General Will (Rousseau) which is the only legitimate expression of the People.

Shouldn't we re-think Ellul's anti-State attitude now that the earnings of the "Web Giants" (Amazon, Facebook, Google, Netflix, LinkedIn) exceed the GNP of many countries? Apple is as rich as Greece and richer than Portugal. If we add to the score their customers—all of us!—the mining of Big Data by the police and Big Corporations to profile populations, and the inhumanity of management practices toward their employees, then isn't the anarchist hippie dream turning into a technocapitalist nightmare? Under these conditions, even anarchists might hope for the State, that coldest of cold monsters, to protect them from the power of these hot new monsters.

Bibliography

Crouch, Colin. *Post-Democracy*. Cambridge, MA: Polity, 2004.
Ellul, Jacques. "Quand les mots perdent leur sens" ("When words lose their meaning"). *Reforme*, n° 73 (August 10, 1946).
Troude-Chastenet, Patrick. "From the 'Contaminated Blood Affair' to the Mediator Scandal: Public Health, Political Responsibility, and Democracy." In *Jacques Ellul and the Technological Society in the 21st Century*, edited by Helena M. Jeronimo, Jose Luis Garcia, and Carl Mitcham, 145–58. Dordrecht: Springer, 2013.
———. "Santé publique et démocratie: l'affaire du Médiator." *Etudes*, Tome 415, n°3 (4153) (September 2011) 185–96.

21

Sham Universe

Notes on the Disappearance of Reality in a World of Hallucinations

BY Doug Hill

> *Doug Hill is a journalist and independent scholar who has studied the history and philosophy of technology for more than twenty-five years. His book,* Not So Fast: Thinking Twice About Technology, *is published by the University of Georgia Press. He blogs at "The Question Concerning Technology" and can be followed on Facebook at @NotSoFastBook, and on Twitter at @DougHill25.*

* * * *

> Identifying the "truth" is complicated.
>
> —MARK ZUCKERBERG, FOUNDER AND CEO OF FACEBOOK[1]

IF EVER A SERIES of events testified to the prophetic vision of Jacques Ellul, the 2016 campaign for the presidency of the United States and the election of Donald J. Trump surely did. In particular, the campaign was flooded by an unprecedented avalanche of "fake news" that uncannily and uncomfortably affirmed Ellul's analysis of propaganda more than fifty years earlier.

This is not to say Ellul could have foreseen the variety and quantity of propaganda that contributed to Trump's victory. Dispatches with little or no regard for the truth were promulgated not only by domestic organizations

---

1. Quoted by Reed, "Does Even Mark Zuckerberg Know."

and individuals with agendas to promote but also by foreign agents impersonating American citizens and by hustlers whose only interest was in making money. After initially denying that it had helped influence the election, Facebook turned over to Congress more than 3,000 election-related ads sponsored by Russian organizations, most of them aimed at fanning the flames of divisive social issues. Other investigations, meanwhile, found that hundreds of Twitter accounts connected to Russia had posted thousands of tweets, many of them produced automatically by bots, attacking Trump's opponent, Hillary Clinton, while Google found that Russian agents had spent "tens of thousands of dollars" buying ads on Gmail and YouTube as well as its search engine. The fact that Trump regularly added his own truth-challenged claims to the mix added to the confusion.[2]

Although Ellul could not have anticipated the scope and scale of this deception—no one could have—I don't think he would have been surprised by its effects. The specifics of technical applications change, but their impacts on human beings remain more or less the same. The essential difference is captured in one of Ellul's favorite aphorisms: A change of quantity often becomes a change in quality.

In this essay, I will review key points in Ellul's discussions of propaganda, drawing mainly on *Propaganda: The Formation of Men's Attitudes*. The chapter's title is taken from *The Technological Society*.[3] My observations will concern what's happening in the United States, although I don't doubt that much of what is happening in the States applies in some fashion to other countries. Revelations concerning the influence of fake news on the campaign continue to appear as I write, so significant ongoing developments will almost certainly not be included in these reflections.

Propaganda as Friend

At the time of Ellul's death in 1994, Google, Facebook, and Twitter did not exist and the Internet in general had not achieved anywhere near the sort of influence it exerts today.[4] Technology enthusiasts have long argued that

---

2. The fake news problem received voluminous press coverage, especially after the campaign. Examples include these *New York Times* reports: Shane and Goel, "Fake Russian Facebook Accounts"; Shane and Isaac, "Facebook to Turn Over"; Shane, "Fake Americans"; and Higgins, McIntire, and Dance, "Inside a Fake News Sausage Factory." See also Dwoskin, Entous, and Timberg, "Google Uncovers"; and Kessler, Hee Lee, and Kelly, "President Trump's List"; in the *Washington Post*.

3. Ellul, *Technological Society*, 372.

4. According to Wikipedia, Google was incorporated in 1998, Facebook was founded as a Harvard website in 2004, and Twitter was launched in 2006. The world

the world wide web would democratize public discourse by offering virtually anyone with a computer connection the opportunity to publicly share his or her ideas; no longer would the gatekeepers of traditional media determine whose voices could be heard. In some circumstances this is true. The Internet and other media can expose us to enlightening, empowering information. However, it has become increasingly obvious that the Internet and other media can expose us to vast amounts of *mis*information, thereby encouraging us to base our opinions and behaviors on distorted perceptions of reality. The role of fake news in the election of Donald Trump has irrevocably affirmed the legitimacy—indeed, the urgency—of that concern.

The question that has been most often raised in response to the fake news issue has been how the major Internet companies can reduce its prevalence on their sites (no one believes it can be eliminated). Ellul, by contrast, devoted much of his attention to explaining why people respond to propaganda as favorably as they do, emphasizing that the pejorative connotation attached to the word "propaganda" obscures how we really feel about it. We think we don't like propaganda, that we don't want to be subjected to it. To the contrary, Ellul said, propaganda achieves the power it has precisely because we so desperately need it. Propaganda helps us maintain our senses of identity and self-worth in an environment in which, thanks to technique, our confidence in those crucial convictions are under constant assault. "There is not just a wicked propagandist at work who sets up means to ensnare the innocent citizen," Ellul wrote. "Rather, there is a citizen who craves propaganda from the bottom of his being and a propagandist who responds to this craving."[5]

Political analysts believe Trump's positions on immigration and trade and his anti-establishment persona appealed to voters who feel they've been displaced by social and economic trends of recent decades. For example, surveys conducted during and after the campaign by the research organization PRRI and *The Atlantic* found that large percentages of white working-class voters "believe American culture and way of life has deteriorated since the 1950s" and that "the U.S. is in danger of losing its culture and identity." Nearly half of working-class Americans responded that "things have changed so much that I often feel like a stranger in my own country."[6]

It's fair to conclude, as Ellul surely would have, that the conspicuous expansion of various applications and ramifications of technique in recent

---

wide web was first released to the public in August of 1991.

5. Ellul, *Propaganda*, 121.

6. Cox et al. "Beyond Economics." See also Gaughan, "5 Things That Explain." Also relevant: Faris et al., "Partisanship, Propaganda, and Disinformation."

decades—globalization, automation, and corporate downsizing, to name a few examples—have contributed substantially to these feelings of displacement and resentment. One of Ellul's central arguments is that human beings are "diminished" by life in the technological society.[7] The stressful conditions in which many of us work; the blighted conditions in which many of us live; the overwhelming pace of change; the constant threats of obsolescence and unemployment; the deadening cascades of information competing for our attention; the impersonality that characterizes our interactions with public and private institutions—all create conditions that undermine our capacities for balance and security. "Never before has the human race as a whole had to exert such effort in its daily labors as it does today as a result of its absorption into the monstrous technical mechanism," Ellul wrote in *The Technological Society*. "It may be said that we live in a universe which is psychologically subversive."[8]

It is hardly surprising that in such a universe, certain audiences would be receptive to the message that they've been cheated out of what is rightfully theirs by a rogue's gallery of scapegoats—politicians, Wall Street bankers, foreign interlopers, politically-correct liberals, and criminal or lazy minorities among them. And despite Trump's labeling of mainstream news outlets such as CNN and the *New York Times* as "fake news," numerous reports have documented that his supporters responded to real fake news (an oxymoron if there ever was one) in vastly greater numbers than did Clinton's supporters.[9] This suggests that Trump's supporters were in general more aggrieved than Clinton's supporters, which would fit Ellul's profile of propaganda's ideal target.

Pre-Selection, Justification, Community

What specific benefits does propaganda offer the beleaguered citizen of the technological society? Most practically, it provides a sorting tool; it tells us what's worth paying attention to. This is a key reason why propaganda has become steadily more important in the era of the Internet. Information is power, we're told, but for most of us wading through the volume of

7. Ellul, *Propaganda*, 149.
8. Ellul, *Technological Society*, 319, 368.
9. See Silverman, "This Analysis Shows," which includes this finding: "Of the 20 top-performing false election stories identified in the analysis, all but three were overtly pro-Donald Trump or anti-Hillary Clinton." One of the *New York Times* articles cited in n. 2 above (Higgins, McIntyre, and Dance, "Inside a Fake News Sausage Factory") documents the same phenomenon in a less empirical manner.

information available today is an overwhelming challenge, one that at some point we simply decline to take on. "It is a fact," Ellul wrote in 1962, "that excessive data do not enlighten the reader or the listener; they drown him. He cannot remember them all, or coordinate them, or understand them; if he does not want to risk losing his mind, he will merely draw a general picture from them. And the more facts supplied, the more simplistic the image."[10] Propaganda web sites, radio, and television programs take advantage of this situation by giving us pre-digested packages of pre-selected information. It may not be comprehensive, balanced, or true, but it's all we have time for.

As pressing as our need for information management might be, there's a far deeper need that propaganda satisfies: the need of individuals living in the technological society for reassurance of their value as human beings. Propaganda offers us an antidote to our diminishment. It tells us that we know things and that what we know matters. That *we* matter. As Ellul put it, propaganda *justifies* us. Bolstered by propaganda, he said, the individual can look down from the heights upon daily trifles, secure in the knowledge that his opinion, once ignored or actively scorned, has become "important and decisive. . . . He marches forward with full assurance of his righteousness."[11]

Obviously human beings have always been prone to confirmation bias—as Paul Simon put it, a man hears what he wants to hear and disregards the rest. But technology has provided us both the ability to immerse ourselves in an all-encompassing confirmatory environment, as well as the motivation to do so. At the same time propaganda platforms can serve as a gathering place for others who feel the same way we do and opportunities to join with them in mutually-reinforcing groups. In a technological environment of alienation and isolation, propaganda can bind us to a community. But these are highly selective rather than diverse communities. They are actively, aggressively disinterested in sharing discussion and views with members of other communities. Again, the point is affirmation, not an exchange of ideas. This leads, Ellul said, to "an increasingly stringent partitioning of our society." The more propaganda there is, he added, "the more partitioning there is."[12]

During the presidential campaign the satirical web site *The Onion* made fun of these conditions with a (fake!) article headlined "Man Forced to Venture Pretty Far Into Wilds of Internet to Have Opinion Confirmed." It began:

---

10. Ellul, *Propaganda*, 87.
11. Ellul, *Propaganda*, 150, 159, 165.
12. Ellul, *Propaganda*, 213.

> Trekking well beyond the comfortable terrain of the first few pages of his Google search, local man Bruce Costas, 35, was reportedly forced to venture deep into the harsh wilds of the internet Wednesday to have his opinion confirmed by outside sources. Costas, who had fervidly espoused the opinion during a conversation earlier in the day, was said to have spent most of his evening slogging through a dense and oftentimes disorienting jungle of uncharted news sites, rarely visited blogs, and broken links in hopes of coming upon some hidden spring of affirmation, however small or isolated, that could corroborate his viewpoint.[13]

The joke was not only that it would be newsworthy if anyone had to look very hard to find like-minded views on the Internet, but also that, if like-minded views *were* hard to find, people would be desperate to find them.

So it is that we live in a time when, despite the availability of unprecedented amounts of information, massive public delusions—climate change denial, the missing Obama birth certificate, the fear that vaccinations promote autism in children, the belief that Saddam Hussein of Iraq was involved in the 9/11 terrorists attacks, to name a few examples—can flourish and successfully resist any attempt at refutation, no matter how well documented. "Effective propaganda needs to give man an all-embracing view of the world," Ellul said. "The point is to show that one travels in the direction of history and progress." This all-embracing view of the world "allows the individual to give the proper classification to all the news items he receives; to exercise a critical judgment, to sharply accentuate certain facts and suppress others, depending on how well they fit into the framework."[14]

Sociological Propaganda

The implications of Ellul's arguments regarding "sociological propaganda" are at least as troubling as his understanding of political propaganda, especially when one considers that a powerful interplay between the two forms played a significant role in the outcome of the 2016 election campaign.

In contrast to propaganda aimed at convincing people on a specific issue, sociological propaganda articulates a much more general collection of beliefs and assumptions that define for an entire society what is considered normal, acceptable, desirable, and beyond question. It is spread

---

13. The *Onion* article is available online at https://local.theonion.com/man-forced-to-venture-pretty-far-into-wilds-of-internet-1819578912.

14. Ellul, *Propaganda*, 146–47.

spontaneously, rather than as a deliberate act, promulgated by lifestyle magazines, advertising, movie stars and pop singers, school teachers, talk-show hosts, preachers, fashion designers, parents, and friends. It speaks out from the products on the shelves of supermarkets and department stores and from the mouths of the people we pass on the street as well as from the styles of their clothes and haircuts. Sociological propaganda produces, Ellul said, "a progressive adaptation to a certain order of things, a certain concept of human relations which unconsciously molds individuals and makes them conform to society." It is, he added, "a sort of persuasion from within."[15]

Sociological propaganda exacerbated the resentments felt by the aggrieved voters who gravitated to Trump. The aggregation of aspirations and mythologies known collectively as "the American Dream" created a set of expectations centered around beliefs that if you work hard and follow the rules you are entitled to a certain degree of security and social position, in addition to a comfortable lifestyle. When those rewards didn't materialize, or when they evaporated, the result was copious anger and a desire to punish those responsible.

Conclusion

Again, Ellul could not have foreseen the massive distribution of fake news or the massive reach of blatantly partisan news platforms that characterized the 2016 Presidential election. The kudzu-like flowering of these poisonous offshoots of the technological tree are completely consistent, however, with one of Ellul's most fundamental convictions regarding technique: that its central motivation is to expand its sphere of influence.

The seemingly eager credulity of Trump's supporters was consistent as well with Ellul's belief that the brutality of the technological society makes affirmation more important than truth. This belief in turn caused him to issue one of the statements that have earned him a reputation for pessimism. "Democracy is based on the concept that man is rational and capable of seeing clearly what is in his own interest," he wrote in *Propaganda*, "but the study of public opinion suggests this is a highly doubtful proposition."[16] It is difficult after the 2016 election not to share those misgivings.

The subject of propaganda stirred in Ellul some of his angriest and least forgiving rhetoric. When we surrender ourselves to propaganda—when we fend off reality in order to reinforce our preconceived ideas of what is and what should be—we are guilty of an ethical failure, he believed. A person

15. Ellul, *Propaganda*, 64.
16. Ellul, *Propaganda*, 124.

who does so is convinced "that he himself, his party, his class, his nation are right, that they represent Good and Justice. It is this conviction that is decisive and which effectively sways man into the field of propaganda." He called such a set of beliefs "autojustification" and condemned them unequivocally. "All ethical behavior seems to me to imply a questioning of self, a reassessment, and the acceptance of one's values being questioned by others," he wrote. "It is the price that must be paid both to measure oneself to the value, and to have a possible relation in truth."[17]

This condemnation is somewhat at odds with Ellul's more compassionate understanding of human diminishment under the lash of technique, but, as Randal Marlin has pointed out, Ellul was always passionate but not always consistent.[18] This points to a challenge. It's hard not to judge Trump's supporters for failing to recognize his multitude of inadequacies, or for ignoring them. Ellul's insistence on the necessity of self-examination, however, applies no matter where we fit on the political spectrum. Technique, he once wrote, doesn't terrorize, it acclimates.[19] Those of us who would honor his legacy must be on guard for ways in which we ourselves may have been acclimated.

Bibliography

Cox, Daniel, et al. "Beyond Economics: Fears of Cultural Displacement Pushed the White Working Class to Trump." *PRRI/The Atlantic Report*, May 2017. https://www.prri.org/research/white-working-class-attitudes-economy-trade-immigration-election-donald-trump/.

Dwoskin, Elizabeth, Adam Entous, and Craig Timberg. "Google uncovers Russian-bought ads on YouTube, Gmail and other platforms." *Washington Post*, October 9, 2017. https://www.washingtonpost.com/news/the-switch/wp/2017/10/09/google-uncovers-russian-bought-ads-on-youtube-gmail-and-other-platforms/?noredirect=on&utm_term=.6cc81b4cef03.

Ellul, Jacques. "The Ethics of Propaganda." *Ellul Forum* 37 (Spring 2006) 3–8. (This article first appeared in *Communication* 6 (1981) 159–75. Translated from the French by D. Raymond Tourville.)

———. *Propaganda: The Formation of Men's Attitudes*. Translated by Konrad Kellen and Jean Lerner. New York: Vintage, 1973.

———. *The Technological Bluff*. Translated by Geoffrey Bromiley. Grand Rapids: Eerdmans, 1990.

———. *The Technological Society*. Translated by John Wilkinson. New York: Knopf, 1964.

---

17. Ellul, "Ethics of Propaganda."
18. Marlin, "Problems in Ellul's Treatment of Propaganda."
19. Ellul, *Technological Bluff*, 388.

Faris, Rob, et al. "Partisanship, Propaganda, and Disinformation: Online Media and the 2016 U.S. Presidential Election." Beckman Klein Center for Internet & Society, Harvard University, August 16, 2017. https://cyber.harvard.edu/publications/2017/08/mediacloud.

Gaughan, Anthony J. "5 Things That Explain Donald Trump's Presidential Election Victory." *U.S. News and World Report*, Nov. 9, 2017. https://www.usnews.com/news/national-news/articles/2016-11-09/5-things-that-explain-donald-trumps-presidential-election-victory. Downloaded October 2, 2017.

Higgins, Andrew, Mike McIntire, and Gabriel J.x. Dance. "Inside a Fake News Sausage Factory: 'This Is All About Income.'" *New York Times*, November 25, 2016. https://www.nytimes.com/2016/11/25/world/europe/fake-news-donald-trump-hillary-clinton-georgia.html.

Kessler, Glenn, Michelle Ye Hee Lee, and Meg Kelly. "President Trump's List of False and Misleading Claims Tops 1,000." *Washington Post*, August 22, 2017. https://www.washingtonpost.com/news/fact-checker/wp/2017/08/22/president-trumps-list-of-false-and-misleading-claims-tops-1000/?utm_term=.64d440e55bbb.

Marlin, Randal. "Problems in Ellul's Treatment of Propaganda." *Ellul Forum* 37 (Spring 2006) 9–10.

Nyhan, "Why the 'Death Panel' Myth Wouldn't Die: Misinformation in the Health Care Reform Debate." *The Forum* 8 (2010) article 5. Available online at: https://www.dartmouth.edu/~nyhan/health-care-misinformation.pdf.

Nyhan, B., et al. "The Hazards of Correcting Myths about Health Care Reform." *Medical Care* 51 (February 2013) 127–32.

Reed, Max. "Does Even Mark Zuckerberg Know What Facebook Is?" *New York Magazine*, October 1, 2017. http://nymag.com/selectall/2017/10/does-even-mark-zuckerberg-know-what-facebook-is.html.

Shane, Scott. "The Fake Americans Russia Created to Influence the Election." *New York Times*, September 7, 2017. https://www.nytimes.com/2017/09/07/us/politics/russia-facebook-twitter-election.html.

Shane, Scott, and Mike Isaac. "Facebook to Turn Over Russian-Linked Ads to Congress." *New York Times*, September 21, 2017. https://www.nytimes.com/2017/09/21/technology/facebook-russian-ads.html.

Shane, Scott, and Vindu Goel. "Fake Russian Facebook Accounts Bought $100,000 in Political Ads." *New York Times*, September 6, 2017. https://www.nytimes.com/2017/09/06/technology/facebook-russian-political-ads.html.

Silverman, Craig. "This Analysis Shows How Viral Fake Election News Stories Outperformed Real News On Facebook." *Buzzfeed*, November 16, 2016. https://www.buzzfeed.com/craigsilverman/viral-fake-election-news-outperformed-real-news-on-facebook?utm_term=.mrWojrry8#.yaPypwwzn.

22

An Insight Into Ellul's Necessary Revolution

BY Adrián Almazán Gómez

> *Adrián Almazán Gómez is a PhD student in philosophy at the Universidad Autónoma de Madrid (Spain). He holds an undergraduate degree in physics and two master's degrees from the same university: one related to physics (nanotechnology and condensed matter) and the other to philosophy (philosophical critique and argumentation). He is a member of the publishing house El Salmón and on the editorial board of the magazine* Cul de Sac.

* * * *

ELLUL'S BOOK *AUTOPSY OF Revolution*, published in 1969, reflects on the revolutionary phenomenon. In this work, Ellul outlines what he would call the "necessary revolution."[1] What would be its characteristics and how applicable is this perspective for today? Ellul's considerations on revolt, revolution, and the difference between them, are a good starting point. The thesis, highly controversial in an intellectual milieu dominated by Marxism, is that we cannot speak of true revolution until the eighteenth century. All revolts and uprisings before that date would be examples of rebellions, qualitatively different from the contemporary concept of revolution.

In Ellul's view these phenomena share a common starting point: a feeling of revulsion, either individual or collective, at an unsustainable situation. Certain social structures are deadlocked and can no longer meet

---

1. In this case, by "necessary revolution" Ellul means the "revolution we need," the desirable revolution. In many of his writings he contrasts "necessity" and "freedom" where "necessity" has a negative connotation of "inescapable," "required," "conditioned"—but that is not its connotation in this discussion.

the demands of the people. A cry for freedom rises against such structures—but it "never describe[s] attempts to create a new order, or the desire for a new form of society or government, or even for the advent of a new freedom."[2] Therefore, a "refusal to advance toward [the] future" prevails.[3] Rather, the rebellion aims at some novelty, excess or deviation that must be corrected in order to return to the old and customary, which is accepted without questioning. For Ellul, it is precisely such refusal to conceive and achieve a new order which defines the recurrent failure of these movements. Once the furor of the revolt is exhausted, the movement fades away (given its rejection of institutionalization), leaving behind only blood and fire and, sometimes, paralysis or suppression of the threats that triggered the movement in the first place.

Everything will change after the eighteenth century revolutionary cycle (England, the United States, and France). These movements will no longer work against history. Instead, revolution will shape the direction of history, ensuring its dynamism. Another essential change, the breaking of the institutionalization taboo, will enable the success of these uprisings. Human revolt is used to take over the State. A fundamental change is thus achieved: revolutions no longer call for freedom against the State. Instead, in Ellul's words, they seek to solve the social question from a position of power: to put an end to all inequalities, to bring social justice, to begin a new era for humanity, etc. State and freedom become thus inseparable, since the former is responsible for human emancipation.[4]

As Ellul points out, after two centuries we find that no revolution has been able to provide a solution to the much-vaunted social question. On the other hand, "a type of constant revolution has existed since 1789. Each successful revolution has left the State enlarged, better organized, more potent, and with wider areas of influence."[5] Finally, from the analysis of different revolutions and theoretical proposals, Ellul concludes that the different phenomena that have been considered revolutions cannot be reduced to just one account or model. There is no historical determinism, no infallible scheme, and no single need in the social world.

---

2. Ellul, *Autopsy*, 24.
3. Ellul, *Autopsy*, 39.
4. Ellul, *Autopsy*, 82–83.
5. Ellul, *Autopsy*, 160.

The Revolution We Need

For Ellul, until the present, revolutions only served to reinforce and strengthen power dynamics, or they were trivialized or deactivated in various ways. So now Ellul wanted to rethink revolution and characterize the kind of revolution we need, the "necessary revolution." "For *revolution* to be *necessary*, two conditions are requisite: first, man must sense to some degree that he cannot endure life as it is, even though he may not be able to explain why. Second, the basic social structures must be blocked, that is, incapable of acting to satisfy express needs or of proving access to that satisfaction."[6]

The necessary revolution starts from the same point as revolt and revolution, i.e. the feeling that a situation is so unsustainable and unacceptable that a break becomes necessary. A characteristic shared with revolt is the reaction "against the predictable course of history."[7] That is the only way to overcome blocked social structures and the aforementioned dynamics. With this concept, he intends to detach from the determinism and narrowness of contemporary revolution theories. Instead he proposes that we should adopt a moral perspective, along the lines of the personalist movement in which he participated. Revolution must be felt as "an ethical command, but a hypothetical one." "[R]evolutionary action implies an ethical choice based on a particular image of man."[8]

However, the very nature of our society acts as an almost insurmountable obstacle to this perspective that places human beings and their moral choice at the core of social dynamics. The reason is that we live immersed in a technological system. The short explanation of this is that human societies have been colonized by technique. Not just machine and production technology but the unquestioned, unlimited rule of efficiency and its values and logic in economics, politics, communications media, and all aspects of human experience.

When Ellul points at technique and the State as blocked structures, he is actually referring to the totalitarian nature of the Technological System. Its spread has rendered obsolete value systems and even human beings as drivers of social action. It has thus condemned the world to be a passive subject of a dynamic deadlock. It is a deadlock because any human aspiration incompatible with its rationality is excluded a priori and prevented from developing. It is dynamic, because such deadlock is based precisely on a continuous growth and extension of such dynamics under the ideology of progress. A

---

6. Ellul, *Autopsy*, 238–39.
7. Ellul, *Autopsy*, 249–50.
8. Ellul, *Autopsy*, 234, 239.

false equation of material and moral progress give rise to the feeling of transformation and change, falsely thought of as "revolution"—but of course in its trivialized sense. Lots of movement but going nowhere.

The Technological System is therefore totalitarian. However, to talk of a certain automatic and autonomous behavior of such system does not mean that its working is alien to human beings. Quite the contrary, the very condition for the above is precisely the transformation and adaptation of human beings, their adjustment, so that they become passive cogs moving at the System's pace. As Ellul puts it

> The technological system contains its own agents of adjustment. Advertising, mass media entertainment, political propaganda, human and public relations—all these things, with superficial divergences, have one single function: to adapt man to technology; to furnish him with psychological satisfactions, motivations that will allow him to live and work efficiently in this universe. The entire mental panorama in which man is situated is produced by technicians and shapes man to a technological universe, the only one reflected toward him by anything represented to him.[9]

Consequently, Ellul identifies as an essential problem the alienation in the form of reification imposed by the Technological System on human beings. Langdon Winner has written that "Ellul's main thesis . . . is that technique is entirely anthropomorphic because human beings have become thoroughly technomorphic."[10]

The issue of technique as a vehicle for the blurring of the human condition is central to any consideration of the possibility of the necessary revolution, since we face an "insoluble dilemma today: things that are perceived as motives for revolt, as tremors of revolution, and that relate to the concrete reality experienced by living persons, have no bearing on necessary revolution, which has a totally different context."[11] This is the main difference between revolution and revolt historically. We cannot rely on our revolts to focus on what must be the targets of the revolution we need, i.e., the core structures of our technological society. Our analysis must single out these structures and show how they frustrate the legitimate needs and aspirations of individuals.

The clear conclusion is that the necessary revolution must be totally at odds with social forces and must be rooted in a refusal. And in a mass

9. Ellul, *Technological System*, 313.
10. Winner, *Autonomous Technology*, 42.
11. Ellul, *Autopsy*, 236.

society, the only true revolutionary foundation is the individual. Neither the working class (taken in by the dynamics of the Technological System), nor the Third World (eager to incorporate the very way of life that would ensure their domestication), nor any other of the candidates for revolutionary leader will do. The necessary revolution today can only be based in the conscious, free individual, the polar opposite of the technological mass.

And Today? Cracks in the System

The assimilative power and depth of reach of the Technological System seem to make almost irreversible the individual subservience to the crowd and the system. Revolution based on the transformation of the individual seems almost impossible. Ellul says as much in his 1972 work *De la révolution aux révoltes* ("from revolution to revolts").[12] There he argued that the general impossibility of individual liberation was equivalent to the impossibility of the necessary revolution in the world.

But how totalizing is the technological system? If we agree with Ellul that no realm, no community or group, can stay outside the totalizing dynamics of the Technological System, the situation must certainly be as dire as he describes. Even Ellul, however, has a more flexible social ontology and distinguishes the System from the society. Socio-ecological conditions can lead to cracks in the system and our range of possibilities widens with new opportunities for a collective transformation. Such socio-ecological conditions are currently in place, since the massive disruption of our social metabolism leads to a fast and profound degradation.

This is a critical point. Despite its holistic logic and its universalizing, almost ineluctable march toward a "brave new world," the technological system is frustrated in its perfectionist, totalitarian implementation. The rationalizing technological system butts up at times against the irrationality and dysfunctionality of the individual and the group and perhaps even the randomness, mutation, and uncertainty of nature itself. Chaos and dysfunction bring with them a degree of openness and change.

With Aurélien Berlan, we could read the developments in the last centuries as a great expropriation.[13] In his view, various impersonal forces (monetization, bureaucratization, and community dissolution)—powerful but not effectively totalitarian like Ellul's Technological System—have pursued their colonization in an open and conflictual way. These forces, sublimated in State and technological structures, have advanced on their

12. Ellul, *De la révolution aux révoltes*, 335–46.
13. See Berlan, *La fabrique des derniers hommes*.

promise of overcoming the inherent limitations of the human condition. In their delusional hubris, such forces have promised an end to illness, complete control of nature, constant progress, the end of labor, and even personal and collective immortality in a future space life. But each advance of these impersonal forces has had its failures and further weakened human and community social structures.

Thus, if with Ellul our goal is to recover individual autonomy vis-à-vis the expropriation carried out by the "blocked structures" of technology and State, our only democratic option would seem to be community organizations located outside of the State and resisting the Technological System. Rather than yielding to the myth of society as an evil to be suppressed by impersonal forces, we should assert the possibility of a maximally unalienated life dependent on a community-based autonomy that consciously and radically rejects the influence of such impersonal forces.[14]

In short, the necessary revolution—the revolution we need today—may need to be ultimately based in the free individual, as Ellul argues. But that free individual, rather than being isolated and crushed in an unmediated encounter with mass society and the technological system, may be best, or only, nurtured and sustained by intermediate, community-based groups, so long as these groups refuse the rule of the nation-state, the technological system, and the values at their core. The necessary revolution, then, must seize the opportunity to restore true community, small groups, neighborhood, and friendship as contexts in which healthy and free individuals might appear.

Bibliography

Berlan, Aurélien. *La fabrique des derniers hommes: retour sur le présent avec Tönnies, Simmel et Weber*. Paris: La Découverte, 2012.

Ellul, Jacques. *Autopsy of Revolution*. Translated by Patricia Wolf. Eugene, OR: Wipf & Stock, 2012.

———. *De la révolution aux révoltes*. Paris: Table ronde, 2011.

———. *The Technological System*. Translated by Joachim Neugroschel. New York: Continuum, 1980.

Weil, Simone. *Réflexions sur les causes de la liberté et de l'oppression sociale*. Paris: Gallimard, 1998.

Winner, Langdon. *Autonomous Technology: Technics-Out-of-Control as a Theme in Political Thought*. Cambridge, MA: MIT Press, 1977.

---

14. See Weil, *Réflexions sur les causes de la liberté et de l'oppression sociale*.

23

Telling a Better Story

Jacques Ellul Against Statist Mythology and For a New World

BY Andy Alexis-Baker

> *Andy Alexis-Baker (PhD, Marquette) is Assistant Professor of Theology and Religious Studies at Arrupe College of Loyola University Chicago. He is the author and editor of numerous books and articles, including,* A Faith Embracing All Creatures: Answering Commonly Asked Questions about Christian Care for Animals *(Cascade, 2012) and* A Faith Encompassing All Creation: Answering Commonly Asked Questions about Christian Care for the Environment *(Cascade 2014). Andy is a member of the IJES board of directors.*

* * * *

ON JANUARY 30, 2016, the Secret Service notified Donald Trump of an imminent threat. Some people at his rally may have planned to throw tomatoes at Trump. So he informed the Iowa crowd, telling them that if they saw any tomato wielding gardeners, "knock the crap out of them. . . . I promise you, I will pay for the legal fees."[1] The crowd cheered. A month later Trump said about a protester at a campaign rally, "I'd like to punch him in the face." He then mused about the good old days when dissenters at a rally would be "carried out on stretchers."[2] The American media condemned Trump's rhetoric, lamenting the downward spiral of American political discourse and the "end of the social contract."[3] Pundits portrayed

1. Fahrenthold and DelReal, "Trump Brags His Way Across Iowa."
2. Corasaniti and Haberman, "Donald Trump on Protester"
3. Zito, "America's Social Contract Is Dead."

Trump's rhetoric as incompatible with American ideals of tolerance and civility. "Whither civility?" they asked.

A year later, Trump ordered the U.S. military to launch Tomahawk missiles into Syria because he assumed, without evidence, that Syria had used chemical weapons on its civilians. Previously, he had already ordered military strikes in Yemen, killing over 1,000 civilians in his first 90 days as President. The American media response to Trump's military action against Syria: unrelenting praise (and complete silence about Yemen). MSNBC's Brian Williams called the missile launch "beautiful."[4] CNN host Fareed Zakaria said, "I think Donald Trump became president of the United States last night."[5] Later, *Newsweek* accused Trump of being America's laziest president ever. *Newsweek* editors suggested he should get to work building a war coalition against Syria.[6] Though millions marched after Trump's election and thousands crammed airports to protest his travel ban on some Muslims, hardly anyone protested Trump's military action against Syria.

In this chapter, I argue that Trump's violence and the clashing responses make sense within the fundamental myths that shape American attitudes to state violence. In the first section, I examine the primary myths that underlie American acceptance of state violence. Ellul did not focus on the social contract myth because he attended to Hegelian political myths that privilege the group over the individual. So Ellul paid more attention to critiquing Marx than liberal democracy and its myths. But it is instructive to use Ellul to critique American individualist mythology because it reveals what is sacred to Americans, and what is sacred organizes our entire lives. In the second section, I will argue that Trump's actions and behavior are consistent with the project of liberalism. In the last part, I tell a different story using Jacques Ellul's reading of biblical texts and suggest some further ways to imagine a new world. The resistance to Trump starts with telling another story.

The Myths We Live By: What Is Myth?

There are different ways to use the term "myth." In common use, myth signals a false story at odds with scientific and historical facts. Most ancient writers, however, did not use "myth" like this, and most modern scholars follow the ancients. Myths are "world-constructing" stories that reveal, in the words of Ellul, "the fundamental image of [the human] condition and the world

---

4. Sullivan, "Media Loved Trump's Show of Military Might."
5. Drum, "Donald Trump Is No More Presidential Today."
6. Nazaryan, "Trump, America's Boy King."

at large."[7] For Ellul, myths tap into human conceptions of the sacred and incite people to action based on an emotional response. Myth "is an image of the most mysterious depths of man in confrontation with a given reality."[8] Through myth, people orient themselves in a sacred world. For example, in the modern world people are bombarded with constant chatter and news so that they become overloaded with information and unable to focus on any one thing for very long at a time. Most of the information remains at a superficial level, masking the reality of our contemporary situation. Myth unifies these fragments of time and space, helping a person and society orient themselves in relationship to what people hold sacred. In this way, myth is "the veritable spinal column of our whole intellectual system."[9] These myths explain everything and give coherence and meaning to life.

The State Myth

For Ellul, myth intertwines with the sacred. But to discover the sacred, we must first work through myths.[10] One of the pervasive myths that explain our disjointed experiences in American culture is that of a social contract. Classical theorists like Thomas Hobbes, John Locke, and Jean-Jacques Rousseau grounded the state on etiological myths that posit an original state of nature in which the primary characteristic is isolated individuals. Hobbes imagines an original state where covetous individuals threaten to kill one another, resulting in constant fear. To overcome continuous warfare, men agree to create a superior power, freeing individuals from fear of each other.[11] John Locke begins differently. Though individuals act how they wish, they act in accord with natural law, which restricts harm, and makes for relative peace. However, warfare arises, particularly over property disputes. Once people commence hostilities, feuds endure because they lack a state to end the warring. For Locke too, violence is the primary reason to abandon the natural condition and create a state.

Notably, social contract myths posit an original disunity between individuals. "The natural liberty of man is to be free from any superior power on earth, and not to be under the will or legislative authority of man, but to have only the law of nature for his rule," writes Locke.[12] Though Rousseau's vision

7. Ellul, *Propaganda*, 116.
8. Ellul, *New Demons*, 91.
9. Ellul, *Presence in the Modern World*, 67.
10. Ellul, *New Demons*, 121.
11. Hobbes, *Leviathan*.
12. Locke, *Two Treatises*, 109, Second Treatise, ch. 4.

of society and the state of nature differs from Locke and Hobbes, Rousseau still posits an original state of radical individualism in which "natural man" forms no lasting relationships, living as solitary hermits. Individual isolation is so complete that adults do not even recognize their mothers.[13]

For Hobbes, Locke, and Rousseau, people form a state to protect their property and lives. Hobbes paints the starkest picture, describing a "war of all against all" so dreadful that people create a "Leviathan" to enforce peace.[14] Locke states that "the pravity of mankind being such that they had rather injuriously prey upon the fruits of other men's labours than take pains to provide for themselves . . . obliges men to enter into society with one another, that by mutual assistance and joint force they may secure unto each other their properties, in the things that contribute to the comfort and happiness of this life."[15] Rousseau claims that individuals form the state to "defend and protect the person and goods of each associate."[16]

Classical social contract myth portrays a prototypical condition comprised of arrant individualism necessitating a state to save individuals from perpetual violence. In the mythology, violence usually means religious violence.[17] To dethrone church powers and set up the state as the supreme power, enlightenment thinkers reimagined religion as an apolitical sphere of church-promoted private beliefs about gods and the afterlife while the state controls the public political sphere.[18] Irrational religion ministers to the individual's private relationship with God. The rational state regulates competing interests to prevent bloodshed. The state becomes a savior offering the means of salvation from chaos.

While the state saves people *from* irrational religious violence, it saves people *for* shopping. Ellul claims that modern people often find meaningfulness through consumerism: "Consumption, along with the technology that produces it and the advertising that expresses it, is no longer a materialistic fact. It has become the meaning of life, the chief sacred, the show of morality, the criterion of existence, the mystery before which one bows."[19] The power to own property and consume products is the outcome of a state within the social contract mythology. The soteriology imagines a

---

13. Modern contractarians also begin from isolated individuals. See Gauthier, *Morals By Agreement*; and Rawls, *Theory of Justice*.
14. Hobbes, *Leviathan*, 85, ch. 13.
15. Locke, *Two Treatises*, 242.
16. Rousseau, *Social Contract*, 49.
17. See Cavanaugh, *Myth of Religious Violence*.
18. On this Fitzgerald, "Negative Liberty," 248–79.
19. Ellul, *New Demons*, 144.

transformed state of nature in which individuals pursue their self-interests through accumulation. This transubstantiation happens magically as the state grows in power.

State violence is, therefore, rational. It is called "force" not violence.[20] The mythology exhibits what Ellul describes as a process of technical reasoning, moving through seven stages: rationality, artificiality, automatism, self-augmentation, monism, universalism, and autonomy.[21] The rationality and artificiality of the reasoning is evident through how the myth subtly pushes the hearer to think of herself as an isolated person apart from the natural or social world. The natural world is not an inescapable sacred domain, but a field of scarce resources over which people compete. The natural social world is outright war. The individual has objectified the natural and social worlds by thinking of herself apart from both and imposing a certain concept of them as pure threats (stripping them of life-giving power, thus desacralizing them). Then hearers are bid to eschew ways of thinking about the world (as a living body, for example) and society (as a body unified though and following Christ, for example) as irrational means to end violence between people who now think of themselves as radical individuals.

The only way to achieve peace in these circumstances is through a rational, secularized state. Everything besides the state becomes an abstraction from reality. Only its violence has redemptive meaning. The police and soldiers become heroes and martyrs as they employ the state's one best way to bring individuals together: war. Social contract mythology demythologizes the natural and social world and explains the state as a new sacred with its own mythology.

A World Created: The Sacred State

The concept of the "sacred" plays an important role in Ellul's analysis of the contemporary world. For him, every society has "an order of feelings, experiences, objects, rites, and words to which people attribute a value that is not directly utilitarian, which they believe to be determinative and independent of their own powers, which they do not think they can reduce to the everyday level or to rationality (like society) but which seem to them to be charged with either a potential of inexpressible energy or an explanatory

20. See Ellul, *Violence*, 3, 4.

21. See Ellul, *Technological Society*, 79–147. In the rest of the paragraph I use David Lovekin's way of describing rationality and artificiality and apply it to social contract myth. For the full technical cycle, see Lovekin, "Technology as the Sacred Order," 203–22.

potential."[22] Moreover, the sacred is an inescapable reality that simultaneously threatens and protects life. It is an imposed order, reimagining time (a cycle of holy days) and space (places of special meaning in the mythology), and from this vantage people distinguish between the sacred and the profane, the permitted and forbidden. The sacred provides a map for how to move through its threat and protection.

The state is a modern sacred. People are unable to imagine a life without the nation-state because it protects, threatens, and gives life order and meaning. It transcends our everyday experience yet is immanent in daily life through its symbols and totems, especially flags and police. The state provides the order of time we experience. The work week and the annual holiday cycle mold our sense of time. The high holy days celebrating the state's salvific violence—President's Day, Memorial Day, Independence Day, Veterans Day, to name a few—mark time with American patriotism and memorial blood sacrifice. History itself is told from the vantage of the state (particularly marking its wars, like ritual sacrifices). Globally, states transform the Earth into a pattern of bordered nation-states. Within the United States the land is divided into towns, counties, and states, all with memorials to the larger United States, spaced for pilgrimage or constant reminder. Thus, the nation-state completely controls time and space, impacting everyday lives. Yet the state transcends the average person's reach, with the capitol buildings of most U.S. states and in Washington D.C. explicitly emulating ecclesiastical structures in Rome, for example. One author calls Washington DC, "a myth in stone."[23] The immanence and transcendence parallels well how religious studies scholars describe the sacred.[24]

So the state, Ellul writes, "is the ultimate value which gives everything its meaning."[25] Fused with notions of nationhood, in which individuals find purpose and meaning as a group, the nation-state becomes "the criterion of good and evil. . . . It is good to lie, kill, and deceive for the nation."[26] This is where the state reaches the apex of technical reasoning. It alone determines life and death and judges right and wrong. Religion's private "ends" do not impact state action. The state becomes the shared end. Therefore, it is immoral not to expand and rely on the state; it is sacred. People kill and die for the nation-state; they break down in ecstatic patriotic frenzy at the national anthem sung just right, and become angry at the slightest insult to the "land

22. Ellul, *Subversion of Christianity*, 52.
23. See Meyer, *Myths in Stone*, 50.
24. See Ellul, *Subversion of Christianity*, 52.
25. Ellul, *Subversion of Christianity*, 80.
26. Ellul, *Subversion of Christianity*, 82.

of the free." To suggest a life without the nation-state makes about as much sense as suggesting to a medieval French person that life without Christ would be good.

In the United States, even free-market capitalist politicians do not attack the state as such. While dismantling environmental, labor, and civil protections and attacking other state agencies, Trump simultaneously increased the military budget in 2017 by sixty-eight billion dollars and hired thousands of new immigration enforcement officers. The argument is solely over which areas of the state to expand most and whether to regulate technologies that harm the environment and other people. However, the growth of technology and populations make the state an absolute modern necessity.

The state is set apart from the quotidian private conflicts so that those who work as state agents can judge and authorize violence to impose order. The state claims to be an arbitrator for resources that the public "many" would decide through violence. The state is therefore a peacemaker. Yet, the state requires citizens to kill and die for it as the sacred institution that makes life tolerable. What seems contradictory is necessary for the state as sacred. It gives and takes life. It bestows meaning through a logic of blood sacrifice. The state is "the god of war and of order."[27] The state offers people security from conflicts that arise as the "many" compete for resources and dominance. The state, therefore, needs conflicts to justify its existence. The state becomes essential for any imaged life within the secular/religious or public/private way of seeing society.

The America that Created Donald Trump

Such is American mythology and the sacredness of the American state. The liberal state's primary doctrines derive from myths about human nature, progress, religion/secular divide, the state and national identity. President Trump offers nothing new. He is a classic liberal whose views of human nature, the state, religion, and freedom align well with the classical theories of liberal democracy. His Muslim travel ban, for example, echoes Locke, who argued that the state cannot tolerate Muslims because they supposedly have allegiance elsewhere.[28] Trump's male chauvinism aligns with classical liberalism's ideal of male rights. They were white male rights, underwriting Trump's racism. His entire campaign was built around the slogan of making "American great" through hearkening back to the deepest myths of American individualism and redemptive violence. He will

27. Ellul, *Subversion of Christianity*, 81.
28. See Locke, *Two Treatises*, 238. Locke also banned Catholics and atheists.

make it great by violently ejecting and killing all who the state deems worthy of such treatment.

American media praised Trump's aggression toward Syria because it fits with the American idea that they are the freest people and seek to spread free market individualism across the world. The United States has routinely used its military for this purpose, starting first with the idea of "manifest destiny." Trump's military action does not conflict with the myths that underlie American views of state power, except that sometimes he has been too critical of American globalization, making neoliberal Democrats uneasy because he was sometimes not warlike enough. The state of nature reappears at the international level as nation-states bully each other with no overarching "state." Statism becomes a form of extreme tribalism (and internationalism merely reinforces statism). It is perfectly reasonable, in statist mythology to drop bombs on Syrians and Yemeni, or threaten North Korea with nuclear holocaust, and at the same time denounce violence in the name of Islam as irrational terrorism.

Trump encourages crowd violence as a theatrical stunt to shock the public, causing nonstop media chatter about him. But there is much more than theatrics at work in Trump's topsy-turvy rhetoric. Such speech scapegoats difference because the protester represents a threat to communal identity and therefore a lapse into the violent state of nature. He must be purged. The protester, or anybody who disagrees with Trump, is no longer in the social contract of that crowd. Trump's violent rhetoric unites his crowd in hostility. That hostility can be toward an immediate person with a protest agenda, a whole class of people like Mexicans or Muslims, or a political opponent. Trump and his crowds reveled in violent rhetoric toward these groups as scapegoats for the perceived ills of America.

State mythology cannot mitigate Trump's individualism. At its best contract theory sees humanity as rights bearers which only help separate people further into isolated individuals. They have no common ends. The best we can hope from the state is that it mitigates conflicts between them, but that state does not launch a genuinely social life together. Ellul writes, "The form which constitutes the group is the opposite of a contract.... The contract, like the group, is very fragile, for nothing is less enduring than the will." Only a sense of transcendence can keep people together in the long run, Ellul argues.[29]

Without a genuinely social way of acting with one another, the best American society can do is go to war and become united in its common effort to defeat some enemy. The state's *raison d'être* is to enforce contracts and

---

29. Ellul, *New Demons*, 53.

protect property, but it cannot unify radically separated and selfish individuals. So periodically war is necessary to discipline Americans and bring them together for a common end: killing enemies. America has no defense against Trump's violence; Trump's violence is America.

A Biblical Tale

So, we need a different story. Ellul helps with this in his books about biblical narratives. In several works, he examines the opening chapters of Genesis, which he believes provide a story to critique our self-understandings by showing us how God sees humanity. Genesis 1 paints a picture of creation in which "everything was truly a whole and the plentitude of God filled everything while wisdom danced before him."[30] God created everything within "the unity of being."[31] Within this plenitude and unity God paused to create humans "within" the image of God (Gen 1:26–27). Ellul thinks that traditional understandings of this image as either individual rational being or microcosms of the universe distort the meaning. Instead, he translates Gen 1:26 as "Let us create humanity within our shadow (of Elohim)," which suggests more of a play on words than an essentialist element within humankind.[32] But Ellul also suggests that the radical vision in Genesis for humanity is one of the relationships that are loving and that what it means to be within God's image is to participate in God's loving unity-plurality.

> God is love. When we spoke about this plurality within the unity of God, we were in effect saying that the only relationship is that of love.... It is in accordance with this vision that all of humanity and all human beings should conduct themselves.... It is therefore not a question of a single man and a single woman, from which the rest of humanity has descended. Adam could be translated as "humanity," understood through this vision of love. It is also not a question of how many people were created, but of this ambiguous being: animal, but one that is unique and an image of God in being love.[33]

The biblical vision clashes with the social contract theory which sees isolated individuals living off on their own and the fear and potential for violence arising anytime they encounter one another. Even at its

---

30. Ellul, "Technique," 128.
31. Ellul, "Technique," 128.
32. Ellul, *On Freedom*, 31.
33. Ellul, *On Freedom*, 30.

best, contract views of humanity see mere rights that separate people into property owners. Ellul points out that God says, "it is not good for man to be alone" (Gen 2:18), which is the direct opposite of the pronouncement God makes on creation in Gen 1: "This is the first 'not good' in creation, 'it' being the solitude of man."[34] The biblical vision, as interpreted by Ellul, is one of unified social life in which the primary mode of interaction is not through violence to take another's property, but the action of love as it participates and reflects God's love and plurality. Moreover, the biblical vision does not see humanity as property owners because the world is not property but God's good creation.[35]

Ellul's vision of God also opposes classical contract theorist theology. For Hobbes, the original state of nature was created by an omnipotent God who commanded Adam to do things through God's "irresistible power," rewarding the obedient and punishing the disobedient.[36] Through reason, humans can recognize God as an eternal, uncaused, immutable, omnipotent, and simple will. But we cannot identify any purpose or end to God's moving the world or what God's nature is. These are mysteries. Here Hobbes posits a conception of God in which sheer will and power lie at the root of God's command. Trump's vision of himself as the boss President fits well with this picture of command as pure will. Ellul's understanding of God is not of a sheer will which commands and punishes. Such a God is abstracted from the narrative depiction and cannot be the God of Israel. The God who creates does not do so from irresistible power but from God's plenitude of love. Though God is one, this God is "so complex that within himself he was a plurality.... This God is so impossible to express and encompasses so many things that he is represented as a plurality."[37] Rather than an arbitrary lawgiver, God, for Ellul, is a nearly inexpressible plurality of overflowing love, whose creation is good and peaceable, not violent and isolating. We can recognize a purpose and meaning to existence within this vision: "humanity is a being launched forward as it were, which tends toward something.... Humanity turns towards love."[38] God is a plurality that overflows with love, which humans are meant to image. This is the first word of Jewish and Christian theology expressed in the myths of Genesis, according to Ellul.

The vision of original unity and peace leads to a distinct view of the state's origins in Ellul's reading of the biblical narrative. The first "state" is

---

34. Ellul, *On Freedom*, 54.
35. Ellul, *On Freedom*, 26–27.
36. Hobbes, *Leviathan*, 237, ch. 31.5.
37. Ellul, *On Freedom*, 20.
38. Ellul, *On Freedom*, 31.

really the first city, which begins civilization. Ellul notes that Cain, the first murderer, built the first city so that he could escape the punishment of nomadic wandering.[39] "The city is the direct consequence of Cain's murderous act and of his refusal to accept God's protection."[40] By building the first city, Cain rejects God and sets out on a path to find security by "struggling against hostile forces, dominating men and nature, taking guarantees that are within his reach, guarantees that appear to him to be genuine, but which in fact protect him from nothing."[41] Cain's descendants further plunge the network of cities into war and plundering the earth. Nimrod, for example, is called a mighty "hunter," but Ellul points out that a good translation of Hebrew might be "plunderer" or "conqueror."[42] "The city is now a center from which war is waged. Urban civilization is warring civilization. Conqueror and builder are no longer distinct."[43]

As the ancient Israelites began to covet and imitate surrounding civilizations, they demanded a king. Ellul remarks that this demand rejects God's love. The entire history of royal power starts with 1 Sam 8 where God tells the Israelite people that their beloved king will draft them into his armies, keep their women barefoot, pregnant, and cooking (such high dreams), and heavily tax them to pay for his wars and luxurious lifestyle. God warns finally: "you shall be his slaves" (1 Sam 8:17). They have rejected God in demanding a king. As Ellul reads it, the history of Israelite kings shows them doing as God warned in 1 Sam 8. The first king, Saul, goes mad. The second king, David, though very pious, commits adultery, murders a friend, and God ultimately rejects him because of his incessant war making (1 Chr 28:3). The third king, Solomon, though incredibly wise, "imposed crushing taxes, built ruinous palaces, and took over 700 wives and 300 concubines! He began to worship other gods besides the God of Israel. He built fortresses over the whole land. When he died, he was hated by everyone."[44] From the most spiritual to the wisest of Israel's kings, they acted against God's peaceable vision and God's call for a new people. Ellul continues by pointing out that throughout the biblical narratives the prophets always stood against royal power and so kings often persecuted and rejected those prophets.[45]

---

39. See Ellul, *Meaning of the City*, 6.
40. Ellul, *Meaning of the City*, 5.
41. Ellul, *Meaning of the City*, 3.
42. Ellul, *Meaning of the City*, 12.
43. Ellul, *Meaning of the City*, 13.
44. Ellul, *Anarchy and Christianity*, 49.
45. Ellul, *Anarchy and Christianity*, 50–52.

What is significant about Ellul's narrative interpretation of the origins of "the state" is that it contrasts starkly with the vision of the modern state as a savior from violence. Rather than saving people from violence, the state condemns people to no other choice but violence. Rather than making peace between warring individuals, the state represents a fracture in the original unity of creation and the creation of powerless individuals before a technique that has no moral bearings beyond itself and its means to power. The state is a technique for enslaving people to itself, not a savior or a liberator.

Rather than a process of rational deliberation, the state is the product of a covetous desire that scapegoats the other, according to Ellul's reading. "Sin is a break with God and all that this entails. When I say that people are not good, I am not adopting a Christian or a moral standpoint. I am saying that their two great characteristics, no matter what their society or education, are covetousness and the desire for power. . . . René Girard has fully shown what the implications of covetousness are."[46] The first city was founded not from rational choice but from sin: the state is based on crime.[47] What people call "law" is based on a sin and cannot be anything but unjust in a world that rejects God.

Notice that Ellul does not begin with pessimism and rejection but with the hopeful vision of peace at the outset of Genesis. Peace is more basic than violence. That vision is one Ellul returns to again and again as he looks for ways to crack the society of technique for a more just social life. But unlike the utopian myths promising salvation and giving humanity a central role in their own salvation, the biblical myths speak of God's action.

The best we can probably hope for from ourselves, Ellul argues, is to witness to the coming peace God already established in Jesus. Ellul believes that Christians ought to advocate for specific laws, institutions, and reforms that address the times in which we live, particularly demythologizing and desacralizing the modern world so that people can question the growth of technology and the state. But "even if society is reorganized along the lines that they have advocated, they must remain in opposition and require yet more, because what God demands is infinite, as is his pardon." This places the Christian in a "permanently revolutionary situation."[48]

Most of all, those who tell the biblical stories "are people who live in expectation . . . people of the future and not of a temporal and natural future, but of the *eschaton*, the future break with the present world."[49] This future

---

46. Ellul, *Anarchy and Christianity*, 20.
47. See also Tilly, "War Making and State Making," 169–87.
48. Ellul, *Presence in the Modern World*, 30.
49. Ellul, *Presence in the Modern World*, 30, 31.

is the work of God alone. Yet, Ellul argues, Christians do not sit passively by awaiting a future but embody that future in the present: "they bear in the present day the elements of the *eschaton*."[50] What are these elements? A good place to start is Isa 11:6–9.

> The wolf shall live with the lamb,
> the leopard shall lie down with the kid,
> the calf and the lion and the fatling together,
> and a little child shall lead them.
> The cow and the bear shall graze,
> their young shall lie down together;
> and the lion shall eat straw like the ox.
> The nursing child shall play over the hole of the asp,
> and the weaned child shall put its hand on the adder's den.
> They will not hurt or destroy
> on all my holy mountain;
> for the earth will be full of the knowledge of the Lord
> as the waters cover the sea.

Isaiah 2:4 states further about this expectation: "they shall beat their swords into plowshares, and their spears into pruning hooks; nation shall not lift up sword against nation, neither shall they learn war any more."

Ellul argues that these expectations of a transformed world await their full realization but are embodied through the way Christians live now, however imperfectly. Jesus and people like Saint Francis of Assisi become the exemplars rather than soldiers, police, and politicians. Their followers, if they really understand who they follow, refuse war. They make peace. They seek justice and peace with the rest of creation. Many of them become so peaceable they refuse to harm intentionally any creature. They do so as a witness to the fullness of the peaceable kingdom; the greatest biblical myth.

The Jewish and Christian peaceable kingdom myth speaks not of humans transforming the world, but of God doing so, and in this myth, we have "the highest expression of the word, it reaches the edge and very limit of the inexpressible, the ineffable, and the unspeakable, as does the divine tetragrammaton."[51] Those who preach this word never preach war like Trump's evangelical advisers, some of whom have blessed his threats of nuclear war. Rather, they call upon world leaders to take a leap of faith into

---

50. Ellul, *Presence in the Modern World*, 31.
51. Ellul, *Humiliation of the Word*, 106.

an unknown future and break with the myths and sacreds of the modern day. These visions are not just for Christians but for all people.

Ellul has been called a pessimist time and again. Yet his vision of what is most fundamental and what will outlast the regimes of the world is anything but pessimism. Naomi Klein ends her book *No Is Not Enough* with a quotation from Jean-Claude Servais: "The hour calls for optimism; we'll save pessimism for better times."[52] I can think of no better statement to describe Ellul's work.

Bibliography

Cavanaugh, William. *The Myth of Religious Violence*. New York: Oxford University Press, 2009.

Corasaniti, Nick, and Maggie Haberman. "Donald Trump on Protester: 'I'd Like to Punch Him in the Face.'" *New York Times*, Feb. 23, 2016.

Drum, Kevin. "Donald Trump Is No More Presidential Today Than He Was Yesterday." *Mother Jones*, April 7, 2017.

Ellul, Jacques. *Anarchy and Christianity*. Translated by Geoffrey Bromiley. Grand Rapids: Eerdmans, 1991.

———. *The Humiliation of the Word*. Translated by Joyce Main Hanks. Grand Rapids: Eerdmans, 1985.

———. *The Meaning of the City*. Translated by Dennis Pardee. Grand Rapids: Eerdmans, 1970.

———. *The New Demons*. Translated by C. Edward Hopkin. New York: Seabury, 1975.

———. *On Freedom, Love, and Power*. Compiled, edited, and translated by Willem H. Vanderburg. Toronto: Toronto University Press, 2010.

———. *Presence in the Modern World*. Translated by Lisa Richmond. Eugene, OR: Cascade, 2016.

———. *Propaganda*. Translated by Konrad Kellen and Jean Lerner. New York: Vintage, 1965.

———. *The Subversion of Christianity*. Translated by Geoffrey Bromiley. Grand Rapids: Eerdmans, 1986.

———. "Technique and the Opening Chapters of Genesis." In *Theology and Technology: Essays in Christian Analysis and Exegesis*, edited and translated by Carl Mitcham and Jim Wrote, 123–37. Lanham, MD: University Press of America, 1984.

———. *The Technological Society*. Translated by John Wilkinson. New York: Vintage, 1965.

———. *Violence: Reflections from a Christian Perspective*. Translated by Cecelia Gaul Kings. New York: Seabury, 1969.

Fahrenthold, David A., and Jose A. DelReal, "Trump Brags His Way Across Iowa on Caucus Day." *The Washington Post*, Feb. 2, 2016.

Fitzgerald, Timothy. "Negative Liberty, Liberal Faith Postulates and World Disorder." In *Religion as a Category of Governance and Sovereignty*, edited by Trevor Stack, Naomi R. Goldenberg, and Timothy Fitzgerald, 248–79. Leiden: Brill, 2015.

---

52. Klein, *No Is Not Enough*, 257.

Gauthier, David. *Morals By Agreement*. New York: Oxford University Press, 1986

Hobbes, Thomas. *Leviathan*. New York: Oxford University Press, 1996.

Klein, Naomi. *No Is Not Enough*. Chicago: Haymarket, 2017.

Locke, John. *Two Treatises of Government and a Letter Concerning Toleration*. New Haven: Yale University Press, 2003.

Lovekin, David. "Technology as the Sacred Order." *Research in Philosophy & Technology* 3 (1980) 203–22.

Meyer, Jeffrey F. *Myths in Stone: Religious Dimensions of Washington D.C.* Berkeley: University of California Press, 2001.

Nazaryan, Alexander. "Trump, America's Boy King: Gold and Television Won't Make America Great Again." *Newsweek*, Aug. 1, 2017.

Rawls, John. *A Theory of Justice*. Rev. ed. Cambridge, MA: Belknap, 1999.

Rousseau, Jean-Jacques. *The Social Contract*. New York: Cambridge University Press, 1997.

Sullivan, Margaret. "The Media Loved Trump's Show of Military Might: Are We Really Doing This Again?" *The Washington Post*, April 8, 2017.

Tilly, Charles. "War Making and State Making as Organized Crime." In *Bringing the State Back In*, edited by Peter Evans, Dietrich Rueschemeyer, and Theda Skocpol, 169–87. Cambridge: Cambridge University Press, 1985.

Tilly, Charles. "War Making and State Making as Organized Crime." In *Bringing the State Back In*, edited by Peter Evans, Dietrich Rueschemeyer, and Theda Skocpol, 169–87. Cambridge: Cambridge University Press, 1985.

Zito, Salena. "America's Social Contract Is Dead." *New York Post*, Oct. 12, 2016.

www.ingramcontent.com/pod-product-compliance
Lightning Source LLC
Chambersburg PA
CBHW050623300426
44112CB00012B/1634